On- and Offline Scheduling of Bidirectional Traffic

vorgelegt von
Dipl. Math. Elisabeth Lübbecke
Prenzlau

Von der Fakultät II – Mathematik und Naturwissenschaften
der Technischen Universität Berlin
zur Erlangung des akademischen Grades

Doktor der Naturwissenschaften
– Dr. rer. nat. –

genehmigte Dissertation

Promotionsausschuss
Vorsitzender: Prof. Dr. Boris Springborn
Berichter: Prof. Dr. Rolf H. Möhring
Prof. Dr. Sven O. Krumke
Prof. Dr. Nicole Megow

Tag der wissenschaftlichen Aussprache: 27. Mai 2015

Berlin 2015
D 83

Bibliografische Information der Deutschen Nationalbibliothek

Die Deutsche Nationalbibliothek verzeichnet diese Publikation in der
Deutschen Nationalbibliografie; detaillierte bibliografische Daten sind
im Internet über http://dnb.d-nb.de abrufbar.

ISBN 978-3-8325-4115-6

Logos Verlag Berlin GmbH
Comeniushof, Gubener Str. 47,
10243 Berlin
Tel.: +49 (0)30 42 85 10 90
Fax: +49 (0)30 42 85 10 92
INTERNET: http://www.logos-verlag.de

Zugl.: Berlin, Technische Universität, Diss., 2015

Cover: Solution calculated by our tool for bidirectional ship
traffic at the Kiel Canal (cf. Chap. 2)

Acknowledgements

After in total more than nine years of COGA I would like to express my gratitude for this long period where it was kind of a second home. In addition to the provided enjoyable working atmosphere COGA constitutes an exciting research group accomplishing impressive work. I wish to thank my adviser Rolf H. Möhring for creating this environment and for giving me the opportunity to be part of it.

This work would not have been possible without the DFG Research Center MATHEON. In addition to the main financial support MATHEON fostered the visibility of Berlin Mathematics that was necessary to initiate the cooperation with the Waterways and Shipping Administration that is the origin of this thesis. I am very grateful to the "Planungsgruppe für den Ausbau des NOK", in particular to Martin Abratis, Ulrich Bösl, Martin Bröcker, and Sönke Meesenburg for their efforts and the delightful collaboration. It was a great time working on that interesting and exciting project of which I learned a lot. I am indebted to Rolf H. Möhring for giving me the trust for this responsibility and a lot of advise and support.

In our group there is a second door that was always open for my concerns. I am very grateful to Nicole Megow for lending me a lot of her time for intensive discussions and helpful advise. I wish to thank her and Sven O. Krumke for taking the effort of the assessment of my thesis.

Sincere thanks go to my coauthors Yann Disser, Shashwat Garg, Max Klimm, Marco E. Lübbecke, Olaf Maurer, Nicole Megow, Rolf H. Möhring, and Andreas Wiese. Discussion is a vital instrument for understanding. I utterly appreciate how much I learned from every single joint work. I thank Felix G. König for stimulating discussions about algorithmic ideas and his support within the Kiel Canal project, Torsten Klug and Thomas Schlechte for our joint considerations to apply their train timetabling approach and our former student assistants Eamonn T. Coughlan, Martin Luy, Mario Meißner and Robert Pankrath for their contributions to implementations.

I thank Felix G. König and Alexander Richter for the pleasant time in our shared office. I am grateful to Torsten Gellert, Wiebke Höhn, Roman Rischke, and Alexander Richter for their remarks improving this document. Special thanks go to Wiebke Höhn for all the useful suggestions, tips and templates and encouraging conversations.

Ganz besonders danke ich auch meinen Familien für ihre Unterstützung, ihr Verständnis und die vielen aufbauenden und motivierenden Worte. Die zusätzliche Arbeitszeit, die mir gerade in den letzten Wochen frei geschaufelt wurde, hat unglaublich geholfen. Die Worte von Allan Borodin und Ran El-Yaniv können kaum treffender sein: „Auch wenn die Zukunft ungewiss ist, kannst du doch immer auf deine Familie zählen" [BEY05].

Berlin, April 2015

Elisabeth Lübbecke

Contents

Contents

Introduction

Who has never waited at a traffic light because all but one lanes of the road were closed for construction works? In this situation, the red light seems to stay for ages until a bunch of few cars from the other side arrive and the lane is finally free for the own passage. The reason is that after the last car from the other side entered the lane we have to wait the complete time the car needs to transit the stretch. The longer the distance the longer is the waiting time. This is a classical example of bidirectional traffic. It is characterized by the property that after one vehicle enters a tight lane, further vehicles moving in the *same* direction can do so with relatively little headway, while traffic in the *opposite* direction usually has to wait until the whole lane is empty again (cf. Figure 1 for a schematic illustration).

Figure 1: Traffic in two directions through a bottleneck. The three vehicles from the right can follow each other passing the tight lane relatively fast, while the two vehicles from the left have to wait a long time until the last vehicle from the right finished the transit of the complete stretch.

For examples like road works, tight mountain passes, or bridges traffic lights are a good solution to prevent collisions of vehicles entering a tight segment at its different ends. But when single-track infrastructures are controlled by some operators the planning process has to deal with the special character of bidirectional traffic, e.g., when coordinating trains on single-track railway lines.

In this thesis we investigate a further prominent example of that kind: the ship traffic at the Kiel Canal. Situated in the north of Germany the Kiel Canal links the North and Baltic Seas. Hence, the canal is operated bidirectionally. Since offshore vessels are not primarily designed for inland navigation, the passing of two ships with large dimensions is not possible at arbitrary positions. To deal with these problems, there are wider areas called sidings within the canal that allow for passing and waiting, cf. Figure 2. Hence, we actually deal with a sequence of bottleneck segments and decisions must be made about who is waiting for whom, where, and for how long. Responsible for these decisions is the Waterways and Shipping Board with a team of nautically experienced expert navigators. They try to distribute necessary waiting times in sidings fairly among all ships.

Since the Kiel Canal has seen a tremendous growth of traffic demand, which is expected to continue, an enlargement of the canal is planned. There are a bunch of possible construction options such as extending or creating sidings or to allow more flexible passing of ships by deepening and/or widening crucial parts of the canal. In order to assess the cost and benefit of these options their combined effects under predicted ship traffic needed to be reliably estimated. In this thesis we present the mathematical and algorithmic foundation

Figure 2: The Kiel Canal with wider sidings that allow for passing and waiting, cf. www.wsa-kiel.wsv.de/Nord-Ostsee-Kanal/Verlauf_u_Querschnitt/.

that leads to an optimization tool which emulates the current ship traffic control. This tool was used to evaluate the various construction options with the aim of selecting a most adequate combination.

A crucial step in the development of our algorithm is the interpretation of bidirectional traffic as a problem related to classical machine scheduling. This interpretation also helps to get theoretical insights for a better understanding of the nature of bidirectional traffic. To do so, we interpret a sequence of segments where dodging and waiting is possible between each two consecutive ones as a linear arrangement of machines similar to a flow shop model, but with the major difference that jobs may run in both directions. Each vehicle is interpreted as a job with two distinct time values: (1) the processing time defining the time spent for entering the segment, i.e., the time between the first entrance of the prow (head, nose) and the moment the stern (tail, rear) has accessed the segment (possibly including necessary security headway), and (2) the transit time specifying the additional time needed to traverse the segment after the entering is finished. While the former prevents the segment from being used by any other job (running in *either* direction), the latter only blocks the segment from being used by jobs running in *opposite* direction. With this compact representation it is possible to generalize some techniques known from the scheduling literature to obtain efficient algorithms with provable performance guarantees. Moreover, we are able to identify properties of bidirectional scheduling that increase the complexity compared to machine scheduling.

The ship traffic at the Kiel Canal constitutes a second interesting issue. While in railway planning the train schedule is determined in advance vessels are allowed to send a transit request through the canal even shortly before their arrival. Hence, it is worth investigating bidirectional scheduling from the online optimization perspective. A popular method for studying the performance of online algorithms not only in scheduling is competitive analysis [Kar+88; ST85]. It provides an effective framework to analyze and classify algorithms based on their worst-case behavior compared to an optimal offline algorithm over an infinite set of input instances. This performance guarantee is called *competitive ratio*. We develop online algorithms for bidirectional scheduling and give upper bounds on their competitive ratios as well as lower bounds on the best possible competitive ratio that any online algorithm can achieve.

Unfortunately, unsatisfactory gaps between upper and lower bounds remain—as for most online problems. A provably optimal online algorithm, w.r.t. competitive analysis, among *all* online algorithms is only known for very restricted problems. For the different machine environments in scheduling, a long sequence of papers emerged in the past two decades introducing new techniques and algorithms to improve upper and lower bounds of competitive ratios. But in particular the given lower bounds stem from *manually* designed input instances. To adopt another course we introduce the concept of *competitive-ratio approximation schemes* that compute algorithms with a competitive ratio that is at most a factor $1 + \varepsilon$ larger than the optimal ratio for any $\varepsilon > 0$. The key ingredient for these schemes is to enumerate over all possibilities that are relevant for adversary and online algorithm.

Outline of the thesis

Chapter 1. This preliminary chapter provides the reader with the foundations necessary for this thesis. After summarizing the relevant standard concepts being applied we continue to consider bidirectional traffic. We give an insight into the everyday life at the Kiel Canal. Having this in mind we introduce different models of bidirectional traffic that are appropriate for distinct purposes. A geometric model of the *Ship Traffic Control Problem (STCP)* is the most precise one describing the feasibility of itineraries in space and time completely. Therefore, it covers all properties of the Kiel Canal that are important when investigating enlargement possibilities. We analyze its combinatorial structure by considering a relaxation formulated as mixed integer program that concentrates on the decisions relevant between consecutive sidings. Essentially, they reduce to the determination of partial orders or say precedences. These observations suggest relations to scheduling. Motivated by this perspective we formulate a third model of a further simplification following terminologies of the scheduling literature. The *Bidirectional Scheduling Problem (BSP)* investigates a set of jobs traveling bidirectionally on a path of segments where the traversal of a job along a segment is governed by its processing and its transit time. We conclude the chapter by an overview of previous work related to ours.

Chapter 2. In the second chapter, we present our method to solve the STCP in the most general variant. The approach reflects the combinatorial and geometric components of the problem as presented in Chapter 1. In addition to the precedence decisions between sidings we need a feasible allotment of parking slots within sidings over time that is consistent. To that end, we suggest the integration of algorithmic ideas from dynamic collision-free routing of automated guided vehicles. We therefore offer a unified view of routing and scheduling which blends simultaneous (global) and sequential (local) solution approaches to allot scarce network resources to a fleet of vehicles in a collision-free manner. By embedding the algorithm into a rolling horizon planning we construct a fast online heuristic. In view of computational experiments on real traffic data expert planners approved that our combinatorial algorithm is well-suited for the demanded decision support. With the help of instance-dependent lower bounds we assess the quality of our solutions which significantly improves upon manual plans.

Chapter 3. The third chapter investigates the bidirectional scheduling problem and its computational complexity in the offline setting. For the special case of one segment and zero transit times, the problem corresponds to total completion time minimization in non-preemptive single-machine scheduling with release dates. From this special case we directly conclude NP-hardness of bidirectional scheduling, cf. Lenstra et al. [Len+77]. Nevertheless, from the application point of view the transit times are dominating the processing times and hence we are especially interested how to deal with delays induced by opposed jobs. Hence, we investigate the simplification where all jobs have equal processing time. NP-hardness results are complemented by restrictions admitting efficient exact algorithms. Therefore, we understand which properties apart from processing times cause difficulties.

Chapter 4. We now switch to the online setting where jobs of an instance of bidirectional scheduling are not known in advance but appear by their release date. Once, an online algorithm has started a job on a segment it is not possible to revert the decision since it corresponds to the movement of the corresponding vehicle. The purpose of this chapter is to provide theoretical insights on the increase of the cost due to the circumstance that vehicles register their transit request only shortly before their arrival. To this end we apply the powerful framework of competitive analysis where the results of an online algorithm over all instances are compared to the optimum an offline algorithm can achieve. We show that there is at least a loss of factor 2 even for a single segment. Restricting additionally to identical jobs there is still a loss of at least 1.6. From above, we bound the possible loss for the general BSP by factor 4 based on a respective competitive online algorithm. To that end, a relatively simple but general idea of Hall et al. [Hal+97] can be applied. For special cases, we concretize the algorithm so that it has polynomial running time yielding approximation algorithms. Identical jobs on a single segment yield symmetries that can be used to get a different online algorithm with a competitive ratio of $(1 + \sqrt{2}) < 2.42$.

Chapter 5. In the fifth chapter, we present the concept of competitive-ratio approximation schemes. To that end, we introduce a new way of designing online algorithms for the example of preemptive parallel machine scheduling. Apart from structuring and simplifying input instances, we find an abstract description of online scheduling algorithms, which allows us to reduce the infinite-size set of all online algorithms to a relevant set of finite size. This is the key for eventually allowing an enumeration scheme that finds an online algorithm with a competitive ratio arbitrarily close to the optimal one. Our method also provides an algorithm to compute the competitive ratio of the designed algorithm, and even the best possible competitive ratio, up to any desired accuracy. We then generalize the ideas to the non-preemptive setting as preparation to ask for a competitive-ratio approximation scheme for bidirectional scheduling.

Chapter 6. The final chapter seeks for almost best possible off- and online algorithms for bidirectional scheduling with respect to performance guarantee. To that end, we ask for approximation schemes and provide for the single segment case both, a polynomial time approximation scheme for the offline setting as well as a competitive-ratio approximation scheme for the online setting. Therefore, we generalize PTAS techniques of Afrati et al. [Afr+99] and the concepts of Chapter 5 from machine scheduling to the bidirectional case.

1

Basics

This preliminary chapter provides the reader with the foundations necessary for this thesis. After summarizing the relevant standard concepts being applied we continue to consider bidirectional traffic. We give an insight into the everyday life at the Kiel Canal. Having this in mind we introduce different models of bidirectional traffic that are appropriate for distinct purposes. A geometric model of the *Ship Traffic Control Problem (STCP)* is the most precise one covering all important details. We analyze its combinatorial structure by considering a relaxation that concentrates on the precedence decisions relevant between consecutive sidings. Motivated by this perspective we introduce the *Bidirectional Scheduling Problem (BSP)*. We conclude the chapter by an overview of previous work related to ours.

1.1 Preliminaries

In this section, we briefly summarize selected research topics, concepts, and notions used within this thesis. However, we do not intend to give an introduction to combinatorial optimization and computational complexity. Instead, we assume the reader to be familiar with common notions as decision and optimization problems, algorithms, running time, asymptotic notation, complexity classes P and NP, etc. and refer to the comprehensive textbook on algorithms of Cormen et al. [Cor+09]. The purpose is in fact to provide a joint foundation by clarifying the used terminology especially from the fields of routing and scheduling and to emphasize the differences of on- and offline optimization.

Graphs. A *directed graph* $G = (V, A)$ is defined as a set of *nodes* V together with a set of *arcs* being a binary relation $A \subseteq V \times V$. An arc $(v, v) \in A$ is called *loop*. Note that each arc $a = (u, v) \in A$ is an ordered pair of nodes leaving its *start node* u and entering its *end node* v and therefore indicating a direction. Formally, we say that a is *incident from* u and *incident to* v and that v is adjacent to u. Additionally, we define two arcs $(u, v) \in A$ and $(v, w) \in A$ to be *incident*.

In contrast, an *undirected graph* $G = (V, E)$ is defined as a set of *vertices* V together with a set of *edges* being unordered pairs $E \subseteq V \times V$ (usually identified with two-element subsets of V). Undirected notions of incidence and adjacency hold for each edge $e = \{u, v\} \in E$. A *complete graph* with n vertices, denoted as K_n, is an undirected graph where any two pairwise distinct vertices are adjacent. A *bipartite graph* is an undirected graph $(V_1 \dot\cup V_2, E)$ with $E \subseteq V_1 \times V_2$, i.e., no two vertices of each $V_i, i = 1, 2$ are adjacent. Respectively, a bipartite graph with $|V_1| = n_1$ and $|V_2| = n_2$ is *complete bipartite* if $E = V_1 \times V_2$ and denoted as K_{n_1,n_2}. A *cut* of an undirected graph G is defined by a partition $V = V_1 \dot\cup V_2$ of the vertex set and consists of the edges $E \cap (V_1 \times V_2)$.

Routing. The term *routing* actually is used in the literature with slightly different meanings. Often, it refers to optimization problems asking for complete tours through a graph minimizing some criteria. A second considered purpose is to connect pairs of nodes by a path through a graph with minimum connection costs. In this thesis, we need the latter interpretation and think of shortest path computation when speaking of routing.

Formally, a *path* in a directed graph $G = (V, A)$ from node $s \in V$ to node $t \in V$ (short an *s-t-path*) is a sequence $P = (v_0, \ldots, v_k)$ of nodes with $s = v_0$, $t = v_k$, and $(v_{i-1}, v_i) \in A$ for $i = 1, \ldots, k$. The *length* of P is defined as $c(P) := \sum_{i=1}^{k} c(v_{i-1}, v_i)$ for a given cost function $c : A \to \mathbb{Q}^+$. (For this thesis, it is sufficient to restrict to cost functions with non-negative values. Therefore, we refrain from discussions on difficulties caused by negative arc costs.)

Shortest path problem

Given: A directed graph $G = (V, A)$, two nodes $s, t \in V$ and a cost
function $c : A \to \mathbb{Q}^+, a \mapsto c(a)$.

Task: Find an *s-t*-path P in G with minimum length $c(P)$.

In the following, we shortly sketch the well-known algorithm of Dijkstra [Dij59] for shortest path computation in graphs with non-negative arc costs. The algorithm maintains two attributes per node v called *labels*: the length $dist_v$ of the current best found path from s to v and the predecessor node $pred_v$ of v on that path. As long as no path from s to v is found the predecessor label of v remains indefinite and the length label of v remains ∞. The nodes are maintained by a priority queue using the length labels as key.

The procedure is described by Algorithm DIJKSTRA. For detailed discussions on shortest path computation and Dijkstra's algorithm we refer to [Cor+09].

---

input: directed graph $G = (V, A)$, nodes $s, t \in A$, cost function $c : A \to \mathbb{Q}^+$
output: shortest s-t-path in G
1 $dist_s := 0$
2 enqueue each $v \in V$ in priority queue PQ
3 **while** PQ is not empty:
4      $u :=$ node with smallest $dist_u$ dequeued from PQ
5   **if** $u = t$:
6   **return** reconstructed path via predecessor labels
7   **foreach** $v \in V$ adjacent to u:
8   **if** $dist_v > dist_u + c(u, v)$:
9              $dist_v := dist_u + c(u, v)$ and $pred_v := u$

Algorithm DIJKSTRA: Compute a shortest s-t-path in a given directed graph G.

We consider now the special setting where the arc costs correspond to transit times. There, the length of a shortest s-t-path yields the earliest possible arrival time at node t

when leaving node s immediately. If some arc cannot be used at a certain time interval (denoted as *forbidden time window*) it might be necessary or better to wait at some intermediate node instead of using a detour. The concept of time-expanded graphs can help. Time-expanded networks are a standard approach when considering flows over time to ensure that arc capacities are respected at any point in time [FF58; FF62; Sku09]. There, every node has a copy at each point in time connected by arcs for waiting. Every original arc has copies in time connecting only those node copies that correspond to a time difference equal to the arcs transit time.

For a formal definition, consider some directed graph $G = (V, A)$ transit times $\tau : A \to \mathbb{N}$ and a time horizon $T \in \mathbb{N}$. The corresponding *time-expanded graph* $G^T = (V^T, A^T)$ has node set $V^T := \{v_\theta \mid v \in V, \theta = 0, \ldots, T-1\}$ and arc set $A^T := A_1^T \dot\cup A_2^T$ with arcs $A_1^T := \{a_\theta = (u_\theta, v_{\theta+\tau(a)}) \mid a = (u, v) \in A, \theta = 0, \ldots, T - 1 - \tau(a)\}$ for transit and $A_2^T := \{(v_\theta, v_{\theta+1}) \mid v \in V, \theta = 0, \ldots, T - 2\}$ for waiting. For each arc of A_2^T we assign a cost value of 1 for waiting and for each arc of A_1^T we assign a cost value equal to the transit time of the corresponding arc in the original graph.

When looking for s-t-paths respecting forbidden time windows of arcs we insert an additional super source connected to each copy of s without additional costs and connect every copy of t with a super target. We delete every arc copy that represents an infeasible entry of the original arc. Using Algorithm DIJKSTRA we can therefore compute a path from the super source to the super target with earliest possible arrival including waiting time. This setting is referred to as *quickest paths* or *quickest route* computation.

Machine scheduling. The term *scheduling* is associated with all kinds of problems that ask for an *optimal allocation of scarce resources to activities over time*, Lawler et al. [Law+93]. This fundamental class of optimization problems spawned a manifold of research and a multitude of literature since the early 1950s, see [Leu04] for a comprehensive collection of examples. Relevant for our purpose is the special area of *machine scheduling*, cf. [Law+93].

A scheduling instance consists of a fixed set of m *machines* $M = \{1, \ldots, m\}$ and a set of *jobs* $J = \{1, \ldots, n\}$. The machines represent the resources that perform the considered activities represented as jobs over time. In a feasible *schedule* no machine can be assigned to more than one job at the same time, and no job is processed by more than one machine concurrently. If more than one machine is considered there are two possibilities: either each job must be processed serially by each machine of a given (ordered) subset $M_j \subseteq M$ or the machines run in parallel and it is immaterial which of the machines executes the jobs.

Each job $j \in J$ comes with a necessary amount of work that must be completed, i.e., its *processing time*. In the case of *parallel identical* machines this amount is given by a value $p_j \in \mathbb{Q}^+$. (The literature also considers *related* or *unrelated* machines with processing times depending on the assigned machines.) In the serial case the amount of requested work is portioned into machine dependent processing times $p_{ij} \in \mathbb{Q}^+, i = 1, \ldots, m$. If $M_j = (i_1, i_2, \ldots, i_{m_j})$ is an ordered tuple, no processing of job j is allowed on machine i_ℓ before the processing on $i_{\ell-1}$ has been completed. Problems with ordered subsets for each job are called *job shop scheduling problems*. The special case of $M_j = (1, 2, \ldots, m)$ for each $j \in J$ is denoted by *flow shop scheduling*. The unordered case is referred to as *open shop scheduling*.

The concrete definition of a schedule assigns jobs for certain time intervals to machines. If the processing of a job on a machine cannot be interrupted once it has been started, we have for each job $j \in J$ exactly one assigned interval $[S_j, C_j)$ with length p_j for the parallel case and exactly one assigned interval $[S_{i,j}, C_{i,j})$ of length $p_{i,j}$ per $i \in M_j$ in the serial case. If otherwise *preemption*, i.e., job interruption, is allowed, multiple assigned intervals per job and machine are possible.

The earliest point in time $C_j \in \mathbb{Q}^+$ where all necessary amount of work of job j has been addressed in a schedule is called its *completion time*. The earliest point in time $r_j \in \mathbb{Q}^+$ job j is available for processing is called its *release date*. Therefore, we say that each job j has a *flow time* of $F_j = C_j - r_j$. Each job j is additionally characterized by its weight $w_j \in \mathbb{Q}^+$. The quality of a schedule can be measured by several optimality criteria depending on the completion times. In our case, the most relevant objective functions are the *total completion time* $\sum C_j$ and the *total weighted completion time* $\sum w_j C_j$. Note that schedules minimizing the total (weighted) completion time also minimize the total (weighted) flow time $\sum F_j$ ($\sum w_j F_j$) and vice versa. For several purposes we also ask for the amount of time that is needed for the complete schedule. Therefore, we denote the latest completion time $C_{\max} := \max_{j \in J} C_j$ as *makespan*. If no job should be scheduled before time R we refer to the difference $|C_{\max} - R|$ as the *length* of the schedule.

Graham et al. [Gra+79] introduced the three-field notation $\alpha \mid \beta \mid \gamma$ as a compact way to specify the considered scheduling variant where α defines the setup, β restrictions to the input, and $\gamma \in \{\sum C_j, \sum F_j, \sum w_j C_j, \sum w_j F_j, C_{\max}\}$ the objective. The setup may be characterized by $\alpha \in \{1, P, Pm, F, Fm, J, Jm\}$ for a single machine, an unbounded number of parallel identical machines, a constant number m of parallel identical machines, and flow shops or job shops with an arbitrary or a constant number of machines. The β field may be empty or a list with a lot of possible entries like r_j for the presence of release dates, *pmtn* if preemption is possible, $p_j = p$ for equal processing times, etc.

We briefly mention two standard scheduling algorithms we want to refer to: The SRPT *rule* for $1 \mid r_j, pmtn \mid \sum C_j$ by Schrage [Sch68] continues each time a new job is released or a running job is completed with an available job having shortest *remaining* processing time, i.e., the difference between its processing time and the amount it was already processed. *Smith's rule* for $1 \mid \mid \sum w_j C_j$ by Smith [Smi56] schedules jobs in non-increasing order of their *Smith ratios* defined as w_j / p_j for each job $j \in J$.

The literature distinguishes between the cases that either all jobs of a scheduling instance are known in advance or not. While optimization algorithms in the *offline* context are able to compute solutions under consideration of the complete instance, algorithms in the *online* setting have to react immediately without knowing the remaining part of the instance.

Offline optimization. The aim of offline optimization is the development of algorithms with three desired properties: (1) the algorithm creates solutions that are feasible, (2) the constructed solutions are optimal, and (3) the algorithm is efficient, i.e., has a running time polynomial in the input length. Algorithms where property (1) and property (2) is guaranteed for any possible instance are called to be *exact*. Algorithms satisfying property (3) for any possible input are called *efficient*.

Recall e.g. the shortest path problem and Algorithm DIJKSTRA. The created solution is a feasible path with minimum costs. Its running time is in $\mathcal{O}(|V| \log |V| + |A|) \leq \mathcal{O}(|V|^2)$

and therefore polynomial in the number of given nodes and arcs. Hence, Algorithm DIJK-STRA is exact and efficient. The time-expanded graph G^T of a given directed graph G for quickest path computation has a size in $\Theta(T|V| + T|A|)$. Hence, its size and therefore the algorithm's running time might not be polynomial in the size of the original graph and the number of unavailability intervals for arcs. Hence, it is exact but not efficient. However, if T is contained in the input as numeric value, the running time is at least pseudo-polynomial. However, as shown by Gawrilow et al. [Gaw+08] there is a polynomial time algorithm solving the described problem optimally.

Also the SRPT- and Smith's rule are shown to be exact. But even small adaptations of the two problems seem to increase the complexity to be intractable: $1 \mid r_j, pmtn \mid \sum w_j C_j$ and $1 \mid r_j \mid \sum w_j C_j$ are both shown to be NP-hard [Lab+84; Len+77]. Even though not yet proven, it is conjectured and the common believe that NP-hard optimization problems do not admit efficient exact algorithms [GJ79]. But still, there are several approaches to deal with problems that are NP-hard or of unknown complexity without claiming to be efficient and exact. The following examples appear in this thesis.

A mathematically strong relaxation of exact algorithms is to insist on properties (1) and (3) and to have still a provable guarantee on the worst-case-performance of an algorithm over all instances instead of property (2). Consider some minimization problem. For an algorithm A and an instance $\mathcal{I}$ of this problem we denote its solution by $\mathsf{A}(\mathcal{I})$. In addition, we denote a solution of an optimal algorithm as $\mathsf{Opt}(\mathcal{I})$. Abusing notation we refer also to the objective value of these solutions by $\mathsf{A}(\mathcal{I})$ and $\mathsf{Opt}(\mathcal{I})$. A is called ρ-*approximation algorithm* if A has polynomial running time and if $\mathsf{A}(\mathcal{I}) \leq \rho\mathsf{Opt}(\mathcal{I})$ for each instance $\mathcal{I}$ of the considered problem [Gar+73; Joh73]. The value $\rho > 0$ is called *approximation factor*. A family $(\mathsf{A}_\varepsilon)_{\varepsilon > 0}$ of $(1 + \varepsilon)$-approximation algorithms is called *polynomial time approximation scheme*. There are several approaches to develop approximation algorithms, see e.g. the comprehensive book of [Vaz01]. One possibility is to base decisions on linear relaxations.

A *linear program* (LP) is an optimization problem with linear objective function defined by linear inequalities

$$(1.1) \qquad \text{minimize} \quad \sum_{j=1}^{n} c_j x_j$$

$$(1.2) \qquad \text{s.t.} \quad \sum_{j=1}^{n} a_{ij} x_j \geq b_i \quad i = 1, \ldots, m$$

$$(1.3) \qquad x_j \geq 0 \quad j = 1, \ldots, n$$

where $a_{ij}, b_i, c_j \in \mathbb{R}$ are given parameters and values for the variables $x_j \in \mathbb{R}^+$ must be assigned such that (1.2) is satisfied and (1.1) is minimized [Chv83]. Linear programs admit efficient exact algorithms, e.g., the ellipsoid method [Grö+81]. If instead of (1.3) each $x_j \in \mathbb{N}$ the resulting optimization problem is called *integer program* (IP). If this is the case for a subset of variables we speak of *mixed integer programs* (MIPs). Without going into the details, there are powerful solution methods, e.g., branch and bound based on enumeration [LD60; Dak65; WN14], implemented in a bunch of generic solvers. These methods allow for instance-wise bounds on the gap between the objective value of achieved solutions and lower bounds on the objective value (even during the solution process). Without running time and memory limitations this yields exact algorithms. For many applications the available (time and space) resources are sufficient or there is an acceptable tradeoff between them and the achieved solution quality. Since lots of NP-hard optimization problems can be formulated as MIPs this is a popular second possibility to deal with them (as alternative to approximation algorithms, depending on the purpose).

Finally, if algorithms are observed to be fast and good enough for many instances they are quiet useful to solve applications, even though no provable statements were made on running time or objective values. Those algorithms are called *heuristics*.

Online optimization. In the online setting, algorithms must take irreversible decisions for iteratively appearing parts of an instance before knowing the remaining parts of the instance. Consider e.g. the example of quickest path computation. While in the offline setting all forbidden time windows are known in advance it might happen in the online case that arcs become unavailable while the vehicle is already on its way and has left node s. It is possible that no quickest s-t-path respecting also the new time window contains the current position of the vehicle. Hence, starting with a quickest s-t-path and adapting the path after a time window appears can yield solutions with higher objective value than the offline optimum. One question in online optimization is if there exists an online algorithm that gives an optimal solution for the complete instance. Such a solution is called *offline optimum* and denoted by Opt, as above. Consider as a second example the online variant of $1 \mid r_j, pmtn \mid \sum C_j$ where each job j becomes known to the scheduling algorithm only at its release date r_j. By this time no interval before r_j assigned to a job can be reverted. Fortunately, scheduling via SRPT satisfies this requirement. Hence, there is an online algorithm that produces optimal offline solutions. In contrast, for the problem $1 \mid r_j \mid \sum C_j$ it can be proven that no such online algorithm exists [Phi+95; HV96].

The example of preemptive single machine scheduling that admits an online algorithm achieving Opt is deemed to be exceptional. Hence, the actual question is which problem specific performance can still be guaranteed for online algorithms. This is the purpose of *competitive analysis*, see the seminal papers by Sleator and Tarjan [ST85] and Karlin et al. [Kar+88] as well as the comprehensive books of Fiat and Woeginger [FW98b] and Borodin and El-Yaniv [BEY05] on online optimization and competitive analysis. It can be considered as online variant of the analysis of approximation algorithms since it provides worst-case guarantees. Since it seeks at first for insights on the best possible performance under the lack of information given unbounded computational resources there are no requirements on the computational complexity of competitive algorithms.

Given a minimization problem, an online algorithm A is called ρ-*competitive* if, for any problem instance $\mathcal{I}$, it achieves a solution of value $A(\mathcal{I}) \leq \rho \cdot \text{Opt}(\mathcal{I})$, where $\text{Opt}(\mathcal{I})$ denotes the value of an optimal offline solution for the same instance $\mathcal{I}$. The *competitive ratio* ρ_A of A is the infimum over all ρ such that A is ρ-competitive. The minimum competitive ratio ρ^* achievable by any online algorithm is called *optimal*.

To prove that ρ is a competitive ratio for a ρ-competitive algorithm examples are given where the ratio achieved by the online algorithm yields a lower bound on the infimum equal to ρ. Moreover, examples for a general lower bound on the competitive ratio that any online algorithm can achieve are used to prove that a competitive ratio is optimal. These worst-case considerations can be considered as a game between an *online player* and a malicious *adversary*. By knowing the reactions of the online player, the adversary aims to construct instances being expensive for the online player but not for the offline optimum. For a general formalization of this game and a general definition of online algorithms we refer to the concept of request-answer-game by Ben-David et al. [BD+94], see also [BEY05, Chapter 7]. The considerations in this thesis are restricted to online scheduling as explained above, i.e., jobs arrive over time and as soon as a job is released all job parameters are known. (For alternative interpretations of online scheduling confer [Sga98].)

Note that competitive analysis is only one possibility to investigate online algorithms. Depending on the respective purpose other methods might be necessary, e.g., for some problem types it is hard to compare different online algorithms under worst-case considerations. We refer to the discussions of Fiat and Woeginger [FW98a] on this topic.

Finally, when online algorithms are actually implemented and utilized to tackle online applications we have to face again the question of running time. In real-time systems, it might even not be sufficient to prove a polynomial running time. The answer on an online request must be calculated fast enough such that the reaction can follow in time. See the detailed discussions of Grötschel et al. [Grö+01] on this issue.

1.2 Background

The Kiel Canal. With more passages than the Panama and Suez Canals together, the Kiel Canal is the world's busiest artificial waterway. It is located in the north of Germany and links the North and Baltic Seas, cf. Figure 1.1. An average of 250 nautical miles (460km) is saved by using the canal instead of the way around Skaw. The Kiel Canal, as the more ecological and safer route, became the basis for the trade between the countries of the baltic area with the rest of the world [Kie14].

Figure 1.1: The Kiel Canal in the northern part of Germany connects the North and the Baltic Seas, cf. `www.wsa-kiel.wsv.de/Nord-Ostsee-Kanal/`.

Since offshore vessels traveling bidirectionally through the canal are not primarily designed for inland navigation, the passing of two ships with large dimensions is not possible at arbitrary positions. This is called a *conflict*. To deal with these conflicts, there are dedicated locations within the canal, called *sidings* or *turnouts*, which are wider to allow for passing and waiting, cf. Figure 2. Decisions must be made about who is waiting for whom, where, and for how long. This traffic control is done by a higher authority, the Waterways and Shipping Board (WSA) with a team of nautically experienced expert navigators. They try to distribute waiting times in sidings fairly among all ships. The objective is to minimize the total transit times of all ships.

Since the end of the 1990s, the Kiel Canal has seen a tremendous growth of traffic demand. Not only the absolute number of vessels is growing, but also the total gross tonnage per ship [Kie14]. As this development is expected to continue, the Kiel Canal may become inoperable when no countermeasures are taken: it is the growing share of large vessels which entail more of the above mentioned conflicts and challenge the navigators. As a remedy, an enlargement of the Kiel Canal is planned, with construction options such as extending or creating sidings or to allow more flexible passing of ships by deepening and/or widening crucial parts of the canal. In order to assess the cost and benefit of these options their combined *effects* under predicted ship traffic needed to be reliably estimated.

The planning context. Any two ships, no matter what size, can pass each other in sidings. Outside sidings, on *transit segments*, legal regulations and nautical parameters precisely define the circumstances under which ships are allowed to meet. Ships are categorized into *traffic groups* 1–6, mainly depending on their dimensions and charge; transit segments are classified into *passage numbers* 6, 7, or 8, reflecting physical dimensions. In simplified terms, two ships may pass each other on a transit segment only if the sum of their traffic group numbers does not exceed the respective passage number, see Figure 1.2. Also, overtaking a moving ship is considered a dangerous maneuver and thus is not an option.

Figure 1.2: A transit segment of passage number 6. The ship of traffic group 4 has to wait in a siding until the ship of traffic group 3 has released the segment (reason: $3 + 4 > 6$). The ship of traffic group 2 can pass (reason: $3 + 2 \leq 6$).

To protect the river bed, velocities of most of the largest ships are limited to 12km/h, that of all others to 15km/h. We assume constant full speed for each ship, as planners do. Because of non-uniform velocities we also need to obey location dependent safety distances.

In each of the 12 sidings situated at the Kiel Canal, in each direction, there is a *passage track* and an additional *waiting track*, see Figure 1.2. A ship travels on the passage track to its waiting position and *instantaneously* changes over to the waiting track where it stays for the full duration of waiting, then moves back to the passage track. In addition to the ship's length safety distances to other ships must be kept, both, when waiting and moving. Overtaking a waiting ship is possible. Under very restricted conditions it is allowed to wait on the waiting track of the *opposite* direction. Obviously, the waiting capacity of a siding is limited, and the choice of waiting positions influences how well this capacity is used. As a rule of thumb, for each ship, the expert planners try not to exceed a waiting time of three hours during the whole journey and one and a half hours individually in each siding. For ships of traffic group 6 this reduces to two hours in total and one hour per siding.

A journey through the canal typically takes seven to nine hours. It may be preempted on many occasions, into side arms or berths. Hence, ships can appear or disappear for planning literally on arbitrary positions within the canal.

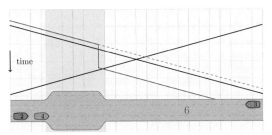

Figure 1.3: Simplified distance-time diagram for the example of Figure 1.2. The x-coordinate corresponds to a position within the canal, the y-coordinate to a point in time. The geographic location of the siding is marked by a darker shading. Each itinerary of the center of a ship is represented by a line plot. Thus, a vertical line within a siding illustrates waiting.

Current traffic control is aided by a *distance-time diagram* (also: time-space or string-line diagram, known from traffic timetabling [Ced07]). Figure 1.3 illustrates an example, a screen shot of the diagram actually in use is given in Figure 1.4.

We finally stress the *online* character of ship traffic control. There is a limited look-ahead, as ships register quite late for a journey through the canal only, about two hours before arrival. Because of a scheduled locking process at the canal's boundaries, arrival times of ships at the first siding are known and preliminary itineraries can be planned. However, these need to be updated upon arrival of new ships.

1.3 The ship traffic control problem

We continue to introduce the new optimization problem of ship traffic control as it occurs at the Kiel Canal. We first define a precise geometric model that completely describes the feasibility of itineraries in space in time. Afterward, we consider a relaxation which ignores capacities in sidings and focuses on the relevant decisions for transit segments. Therefore, it helps to understand the combinatorial structure of the problem.

1.3.1 A precise geometric model

A *canal* is represented as an interval $C \subset \mathbb{R}$, partitioned into a set $\mathcal{E}$ of m intervals, called *segments*, i.e., $\cup_{E \in \mathcal{E}} E = C$. Elements of a distinguished subset $\mathcal{T} \subset \mathcal{E}$ are called *sidings*; we refer to all other segments as *transit segments*. A set of *ships* $S = \{1, \dots, n\}$ induces a set of requests $R = \{(s_i, t_i, v_i, r_i, h_i) \mid i \in S\}$ with *start* and *target* $s_i \neq t_i \in C$ at arbitrary positions in the canal, a *velocity* v_i, a *release time* r_i, and a *parking distance* h_i (see below). The travel direction of a ship (its *heading*), or simply the ship $i \in S$ itself

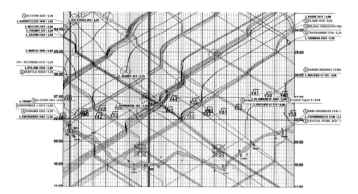

Figure 1.4: Screen shot of the distance-time diagram which is used for traffic control by the Waterways and Shipping Board. The lower part below the current time of 08:00 allows for an interactive traffic control in the near future, like signaling ships to wait in sidings etc. Straight line itineraries reflect future plans.

is called *rightbound* if $t_i > s_i$ and *leftbound* otherwise. Two ships heading in the same direction are called *aligned*, otherwise *opposed*. The velocity of a ship $i \in S$ and the length of a segment $E \in \mathcal{E}$ define the transit time τ_{iE} of ship i along segment E. For each pair of ships $(i, j) \in S \times S$ and each segment $E \in \mathcal{E}$ we are given a *vertical distance* $v_{ijE} \geq 0$. It ensures a sufficient headway between two (aligned or opposed) ships, as defined by the sizes of the ships and the required safety distance (translated to the corresponding transit time via the velocities). Since the safety distance between two aligned ships can depend on the size of the ship following behind, $v_{ijE} \neq v_{jiE}$ is possible. The *horizontal parking distance* $h_i \geq 0$ of each ship $i \in S$ defines the required space when the ship is waiting in a siding. Figure 1.5 illustrates a small problem instance.

For each ship, decisions must be taken concerning space within the canal and time, simultaneously. We call this finding *dynamic routes*. When speaking of a ship's position we actually refer to the ship's center, without further notice. A *solution* is given by a set of simple polygonal chains in $C \times \mathbb{R}$, or *polylines*, one for each request, $\mathcal{P} = \{P_i = ((p_{i,1}, \tau_{i,1}), \ldots, (p_{i,k_i}, \tau_{i,k_i})) \mid i \in S\}$, such that the following properties hold. For each ship $i \in S$, polyline P_i defines a dynamic route from the ship's start position and release time to its target position, i.e., $p_{i,1} = s_i$, $\tau_{i,1} = r_i$ and $p_{i,k_i} = t_i$. The slopes are defined by the ship's velocity, the slope's sign reflects the ship's heading, and waiting (zero velocity, infinite slope) is allowed only in sidings, i.e., either $\frac{|p_{i,\kappa} - p_{i,\kappa-1}|}{\tau_{i,\kappa} - \tau_{i,\kappa-1}} = v_i$ with $(p_{i,\kappa} - p_{i,\kappa-1})(t_i - s_i) > 0$ or $p_{i,\kappa} = p_{i,\kappa-1} \in \cup_{T \in \mathcal{T}} T$ with $\tau_{i,\kappa} > \tau_{i,\kappa-1}$ for each breakpoint $\kappa = 2, \ldots, k_i$. A ship is allowed to wait in each siding at most once, i.e., a polyline's number of breakpoints per segment is bounded by 2. The *waiting time* of ship $i \in S$ in a solution $\mathcal{P}$ is defined as the difference of i's actual passage time in $\mathcal{P}$ and i's minimum possible passage time, i.e., $W_i := (\tau_{i,k_i} - \tau_{i,1}) - |t_i - s_i|/v_i$. A *feasible* (or synonymously *collision-free*) solution $\mathcal{P}$ satisfies the following two properties, cf. Figure 1.6.

Passing rules: Line segments corresponding to two *moving* ships $i, j \in S$ (i.e., two seg-

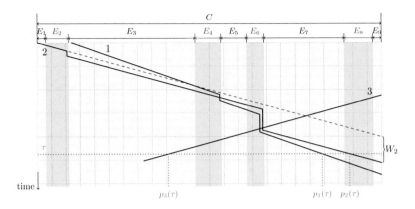

Figure 1.5: A problem instance together with a feasible solution. The partition of the canal C into segments $E_1, \ldots, E_9$ is shown at the top. The shaded segments correspond to the subset $\mathcal{T} = \{E_2, E_4, E_6, E_8\}$ of sidings. The three solid polygonal chains ("polylines") represent three ships navigating the canal. The start point of each polyline is given by the corresponding ship's start position and release time; the velocity defines the slope. By a dashed line we signify a ship's theoretical onward journey if there were no constraints which enforced waiting. For ship 2 it shows how the waiting time w_2 is calculated. The dotted lines illustrate how positions of ships are translated into a given point in time by the functions $p_i(.)$.

ments of finite slope) are not allowed to be too close to each other (or even to cross) if i and j are aligned (in particular, no overtaking) or if i and j are opposed and their passing is not allowed at the respective position. In both cases this will be ensured by the minimal vertical distance v_{ijE} between the line segments.

Parking distance: The required space of a waiting ship defined by the horizontal parking distance h_i must lie completely within the siding. Furthermore, this space must not be occupied by a second ship which is waiting concurrently.

In order to express feasibility mathematically, we need the following definitions. The map $p_i : [\tau_{i,1}, \tau_{i,k_i}] \to C$ assigns each point in time $\tau \in [\tau_{i,1}, \tau_{i,k_i}]$ to the corresponding position of ship $i \in S$ within the canal at time τ as defined by the polyline P_i. For each ship position $p \in C$, the inverse $p_i^{-1}(p) =: \tau_i(p)$ gives a time interval, probably empty or a single point. The open set $H_i(p) := \{[p - h_i, p + h_i] \times \tau_i(p)\}^\circ$, which is the interior of the rectangle occupied for potential waiting at position p, is non-empty only if waiting actually takes place at point p. Let $\overline{\tau_i}(p) := \max \tau_i(p)$ denote the latest point in time at which ship i stays at position p, and similarly $\underline{\tau_i}(p) := \min \tau_i(p)$ denotes the earliest point in time, respectively. Abusing notation, we will reference the interval $[\min\{a,b\}, \max\{b,a\}] \subset C$ by $[a, b]$ when both, $a \leq b$ or $b \leq a$ is possible, depending on the currently considered heading. Mathematically, a solution is *feasible* or *collision-free* if the following holds:

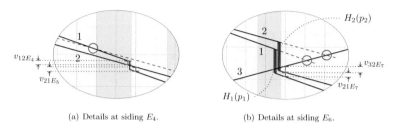

(a) Details at siding E_4.　　　　　　　　(b) Details at siding E_6.

Figure 1.6: Details of Figure 1.5 to illustrate collision-freeness. In part (a) it is not allowed for the faster ship 2 to overtake the slower ship 1 outside sidings. The circle indicates the conflict which would result from violating the required positive vertical distance. To avoid this, ship 2 has to wait in siding E_2 so that ship 1 can wait in siding E_4 to let it pass. There, the required vertical distances indicated by small intervals are respected. Likewise, for conflicting opposed ships in part (b), positive vertical distances forbid the crossing of the respective line segments. Also in part (b), the required space for waiting of ship 1 and 2 in the siding is depicted by the rectangles which have to be disjoint.

Passing rules: For all $i \neq j \in S$, $E \in \mathcal{E}$ and $p \in E \cap [s_i, t_i] \cap [s_j, t_j]$ with $\overline{\tau_i}(p) \leq \underline{\tau_j}(p)$:

$$\underline{\tau_j}(p) - \overline{\tau_i}(p) \geq v_{ijE} \ .$$

Parking distance: For all aligned ships $i \neq j \in S$, $T \in \mathcal{T}$ and $p_1, p_2 \in T$:

$$H_i(p_1) \subset T \times \mathbb{R} \qquad \text{and} \qquad H_i(p_1) \cap H_j(p_2) = \emptyset \ .$$

Due to its geometric character we synonymously refer to a feasible solution also as *geometry*. The considered objective is the minimization of the *total waiting time* $\sum_{i \in S} W_i$.

Ship Traffic Control Problem (STCP)

Given: A canal C and a set of ships $S = \{1, \ldots, n\}$ together with a set of travel requests $R = \{(s_i, t_i, v_i, r_i, h_i) \mid i \in S\}$.

Task: Find a feasible solution $\mathcal{P}$ minimizing $\sum W_i$.

Remark. Our model of parking in sidings is stronger than simply requiring that the length of a siding is not exceeded by parking ships, at no point in time. We refer to this weaker model as *knapsack relaxation* (one may assume that all ships wait at the end of a siding). Our stronger model is much related to orthogonal *rectangle packing* on a strip [Bak+80]. If all ship dimensions are identical the weaker knapsack condition suffices to ensure a corresponding rectangle packing. In general, it is strongly NP-hard to decide if a feasible solution to the knapsack relaxation implies a corresponding feasible rectangle packing; see [Błą+13; DVDS06; Gün+14] for discussions and proofs.

In what follows, when speaking about time, it helps to define a common reference point. To this end, consider a ship $i \in S$ at position p_1 at time τ_1 and determine for some position p_2 the corresponding point in time τ_2 on the straight line through (p_1, τ_1) with slope defined by the heading and velocity of i (e.g., the dashed lines in Figures 1.5 and 1.6). Formally, given a time value τ_1 w.r.t. position p_1 for a rightbound (resp. leftbound) ship i we define $\tau_2 := \tau_1 + (p_2 - p_1)/v_i$ (resp. $\tau_1 - (p_2 - p_1)/v_i$) to be the corresponding time *translated to position p_2*.

1.3.2 Scheduling on transit segments

Having discussed the problem's geometry in the previous section, we now focus on its combinatorics. We first consider a relaxation which ignores all constraints in sidings, in particular capacities, and thus concentrates on the combinatorial decisions to be taken on transit segments. Throughout this section, we assume that all parking distances and all vertical distances for sidings are zero, i.e., $v_{ijT} = 0$ for all $i, j \in S$ and $T \in \mathcal{T}$, and $h_i = 0$ for all $i \in S$. We refer to it as *combinatorial relaxation*.

Ignoring all constraints in sidings, we can derive a solution just from the *time* a ship enters each segment, as each waiting position becomes feasible. A mixed integer program (MIP) will help us describe the relaxation, see Figure 1.7. Let E_i^+ denote the segment a ship must pass after segment $E \in \mathcal{E}$ when heading in the same direction as ship $i \in S$. *Visit time variables* d_{iE} specify when ship $i \in S$ enters segment $E \in \mathcal{E}$. On transit segments, visit time variables are linked by the transit time τ_{iE} in Equation (1.5). On siding segments $T \in \mathcal{T}$, waiting is allowed which is reflected by *waiting time variables* $w_{iT} \geq 0$ for each ship $i \in S$. Equation (1.6) links visit and waiting times in sidings. Lower bounds (1.9) on visit times enforce release times.

$$
\begin{array}{lll}
(1.4) & \text{minimize} & \displaystyle\sum_{i \in S,\, T \in \mathcal{T}} w_{iT} \\[2ex]
(1.5) & \text{s.t.} & d_{iE} + \tau_{iE} = d_{iE_i^+} & \forall i \in S,\ E \in \mathcal{E} \setminus \mathcal{T} \\[1ex]
(1.6) & & d_{iT} + \tau_{iT} + w_{iT} = d_{iT_i^+} & \forall i \in S,\ T \in \mathcal{T} \\[1ex]
(1.7) & z_{ijE} = 1 \ \Rightarrow\ & d_{iE} + \Delta(i,j,E) \leq d_{jE} & \forall E \in \mathcal{E} \setminus \mathcal{T},\ (i,j) \in C_E \\[1ex]
(1.8) & z_{ijE} = 0 \ \Rightarrow\ & d_{jE} + \Delta(j,i,E) \leq d_{iE} & \forall E \in \mathcal{E} \setminus \mathcal{T},\ (i,j) \in C_E \\[1ex]
(1.9) & & \underline{d}_{iE} \leq d_{iE} & \forall i \in S,\ E \in \mathcal{E} \\[1ex]
(1.10) & & 0 \leq w_{iT} & \forall i \in S,\ T \in \mathcal{T} \\[1ex]
(1.11) & & z_{ijE} \in \{0,1\} & \forall E \in \mathcal{E} \setminus \mathcal{T},\ (i,j) \in C_E
\end{array}
$$

Figure 1.7: Mixed integer program for a combinatorial relaxation where all feasibility conditions concerning passing and waiting in sidings are relaxed. Constraints (1.7) and (1.8) can be linearized using a "big M" formulation.

Complications arise when ships i and j are not allowed to pass (when opposed) or overtake (when aligned) each other on segment E, i.e., they have a *conflict* on E (this is equivalent to $v_{ijE} > 0$). We say that ship i *precedes* ship j if ship i passes *each position* of segment E before ship j, i.e., $\overline{\tau_i}(p) < \underline{\tau_j}(p)$ for all $p \in E \cap [s_i, t_i] \cap [s_j, t_j]$. For a conflicting pair, it must be decided, which ship precedes the other. Figure 1.8 shows that precedence can be quite involved. Define $\Delta(i,j,E)$ (or simply $\Delta(i,j)$ if the segment is obvious) to be the smallest value such that the condition

(1.12)
$$d_{iE} + \Delta(i, j, E) \le d_{jE}$$

ensures the required vertical distance v_{ijE} when ship i precedes ship j on segment E. The Δ need not be symmetric, see Figures 1.8(b) and 1.8(c). Figure 1.8(c) also demonstrates that the visit time d_{iE} of the preceding ship i can even be *later* than the visit time d_{jE} of the succeeding ship j. Hence, the Δ values can be negative. Also in this case,

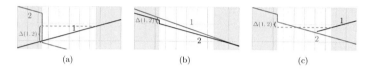

| (a) | (b) | (c) |

Figure 1.8: Examples for calculating Δ: Since ship 1 of Figure (b) is slower than ship 2, $\Delta(1, 2) > \Delta(2, 1)$. Special attention is necessary when a request starts or ends on a segment: This also yields asymmetric Δ values that can even be negative, see Figure (c).

Denote by $C_E \subseteq S \times S$ the set of all conflicting pairs of ships on E. Depending on the decision which ship precedes on a segment, either Condition (1.12) must hold as stated, or with the roles of ships i and j reversed. This *precedence or scheduling decision* is represented by a binary variable $z_{i,j,E}$ and Implications (1.7) and (1.8) precisely reflect the above dichotomy. An assignment of the z variables defines for each transit segment $E \in \mathcal{E} \setminus \mathcal{T}$ a partial order $\zeta(E) \subset S \times S$ on the set of ships where $(i, j) \in \zeta(E)$ if and only if $(i, j) \in C_E$ and ship i precedes ship j on segment E, i.e., $z_{ijE} = 1$. We refer to the collection (not the union) $\{\zeta(E) \mid E \in \mathcal{E} \setminus \mathcal{T}\} := \zeta$ of these partial orders as the *combinatorial frame* of a solution or its *combinatorics* for short.

The determination of the combinatorics is the hard part of this relaxation: Once the precedences are decided on all segments, i.e., z_{ijE} variables are fixed in MIP (1.4)–(1.11), an optimal assignment of the visit time variables (a geometry) can easily be obtained by solving the resulting linear program, or proven to be infeasible if no assignment exists. In the former case, we refer to each corresponding geometry as *realization* of the given combinatorics and call the combinatorics itself *realizable*. Due to the observation that orders imply time values in the combinatorial relaxation, we also employ the term *schedule* when referring to combinatorics, slightly abusing the exact notion.

Remark. The LP relaxation of MIP (1.4)–(1.11) is very weak. Since $z_{ijE} = 0.5$ for all i, j, E is a feasible fractional solution, one obtains the trivial lower bound of zero. The model is still well suited to understand the combinatorial structure of the problem and good enough for first experimental comparisons. We therefore do not discuss models with stronger relaxations here.

Remark. The presented scheduling interpretation is also common in train timetabling on single tracks. Hence, models with similar underlying idea are used, we refer exemplarily to [Szp73; CL95; Lus+11, Sec. 2]. Nevertheless, there is the difference that the passing of only a subset of ship pairs is forbidden and that ships can start and end virtually everywhere in the canal. This entails a variety of configurations of precedence, see again

Figure 1.8. Our introduction of the conflict sets and the Δ values enables us to give a compact formulation by using just one unified statement, namely Condition (1.12), to ensure precedence between conflicting ships. Moreover, this gives us a way to consider only a segment's boundary to ensure precedence *on the whole segment*. Note that we can simply define visit time variables d_{iE} for each ship $i \in S$ and for each segment $E \in \mathcal{E}$ even if the start point of E is not included in the request interval $[s_i, t_i]$. This is done via translation of the time values between the interval boundaries and the start point of E (w.r.t. i's velocity and heading). That is, we think of each ship as traversing the whole canal, even though conditions concerning conflicts are active precisely on the ship's request interval only.

1.4 Bidirectional scheduling

Motivated by the observations on the combinatorial structure of the STCP we develop in the following a third model referring to the terminologies from the scheduling context with the goal to get theoretical insights on the character of bidirectional traffic. To that end, we even further simplify the considered problem by totally ignoring the transit through sidings to completely concentrate on the decisions therebetween.

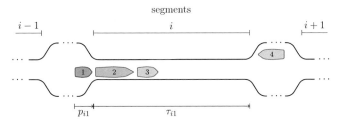

Figure 1.9: Bidirectional scheduling of ship traffic. The processing time p_{ij} of job j is the time needed to enter segment i with sufficient security headway, i.e., the delay before following jobs in the same direction may enter the segment after j. The transit time τ_{ij} is the time j needs to traverse the entire segment i once entered. Job 4 has to wait until jobs $3, 2$ and finally 1 have left segment i. The time to cross sidings is neglected.

Intuitively, we imagine our setting as a path of consecutive machines—or say segments—connected by nodes representing sidings for waiting with neglected length. Confer Figure 1.9 with the following. Vehicles traveling along this path in both directions are considered as jobs, each with a release date and a designated start and destination node. The movement of each job j along a segment i is governed by two quantities: its *processing time* p_{ij} and its *transit time* τ_{ij}. The processing time p_{ij} of a job j is the time needed for the corresponding vehicle to fully enter segment i (including sufficient security headway), and the transit time τ_{ij} is the additional time needed by the tail of the vehicle to traverse segment i (until the entire vehicle left the segment). While a vehicle is entering a segment of the path, no other vehicle may do so. The next vehicle in the same direction can enter immediately afterward, whereas vehicles in opposite direction have to wait until the segment is empty again in order to prevent a collision. Consequently, the processing time of a job prevents the segment from being used by any other job (running in *either* direction), while the transit time of a job only blocks the segment from being used by jobs running in *opposite* direction.

Fixing the points in time when the jobs start entering the distinct segments defines a schedule for bidirectional traffic. An example with two segments and four jobs is illustrated in Figure 1.10. The presentation is guided by the design of distance-time-diagrams. Job movements are represented by parallelograms of the same color. The processing time of a job on a segment is reflected by the height of the corresponding parallelogram, while the transit time is the remaining time (y-distance) to the lowest point of the parallelogram. For comparison, trajectories of our distance-time-diagram representing the movement of the vehicles center would now be situated within the parallelograms. In a feasible schedule, job parallelograms may not overlap, and, in particular, a job can only begin being processed at a segment once it has fully exited the previous segment. The latter restriction relies on the assumption that segments are not necessarily connected geographically. Our results are also applicable if entering of a succeeding segment can start while the preceding segment is left. Nevertheless, we use the other model due to ease of exposition and consistency with flow or job shop scheduling.

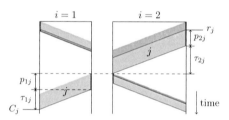

Figure 1.10: Representation of a bidirectional schedule on two segments ($i = 1, 2$) with four jobs as a distance-time-diagram. In this example, all jobs are processed immediately at their release date. Job j is released at time r_j at the right end of segment 2 and needs to reach the left end of segment 1. Since it never has to wait, its completion time is smallest possible: $C_j = r_j + p_{2j} + \tau_{2j} + p_{1j} + \tau_{1j}$. Note that with a min sum objective it makes sense for the two rightbound jobs to switch order while waiting at the central node.

The processing time of some job j may vary among the segments if the vehicles velocity changes during its journey. Since we assume constant speed for each vehicle it suffices to consider the restriction of $p_{ij} = p_j$. If we assume on the other hand that all vehicles travel with unit speed we can furthermore restrict to the case of $\tau_{ij} = \tau_i$.

Formally, we are given in the *bidirectional scheduling problem* a set $M = \{1, \ldots, m\}$ of segments which we imagine to be ordered from left to right. Further, we are given two disjoint sets of J^r and J^l of *rightbound* and *leftbound* jobs, respectively, with $J = J^r \cup J^l$ and $n = |J|$. As in Section 1.3.1, we use the notions of *heading*, *aligned* and *opposed*. We also denote the heading of each job j by $d_j \in \{r,l\}$. Each job is associated with a *release date* $r_j \in \mathbb{Q}^+$, a *weight* $w_j \in \mathbb{Q}^+$, a *start segment* s_j and a *target segment* t_j, where $s_j \leq t_j$ for rightbound jobs and $s_j \geq t_j$ for leftbound jobs. A rightbound job j needs to cross the segments $s_j, s_j + 1, \ldots, t_j - 1, t_j$, and a leftbound job needs to cross the segments $s_j, s_j - 1, \ldots, t_j + 1, t_j$. We denote by M_j the set of segments that job j needs to cross. Each job j is associated with a processing time $p_j \in \mathbb{Q}^+$ and each segment i is associated with a transit time $\tau_i \in \mathbb{Q}^+$. We call $p_j + \tau_i$ the *running time* of job j on segment i.

We address the circumstance that some opposed vehicles are able to pass each other, in the most general way by a bipartite compatibility graph for each segment. There, vertices correspond to jobs and two opposed jobs are connected by an edge if they can cross the segment concurrently. Formally, for each segment i, we are given a bipartite *compatibility graph* $G_i = (J^r \dot\cup J^l, E_i)$ with $E_i \subseteq J^r \times J^l$. Two jobs j, j' that are connected by an edge in G_i are allowed to run on segment i concurrently and therefore called *compatible on i*. Otherwise they are *incompatible* on i or have a conflict. Two aligned jobs j_1, j_2 are defined to have the same *compatibility type* if the set of jobs compatible with j_1 is equal to the set of jobs compatible with j_2 on each segment, i.e., if $\{j : \{j_1, j\} \in E_i\} = \{j : \{j_2, j\} \in E_i\}$ for the compatibility graphs G_i of each segment i. In the special case of one compatibility type per heading we speak of *uniform compatibilities*, i.e., either all pairwise opposed jobs are compatible or none—depending on the segment. Abusing notation, we denote a compatibility graph for the latter case by $G_i = \emptyset$.

Under the assumption of unit speed among the vehicles we give the following formal definition of a feasible solution. A *schedule* is defined by fixing the start times S_{ij} for each job j on each segment $i \in M_j$. The *completion time* of job j on segment i is then defined as $C_{ij} = S_{ij} + p_j + \tau_i$. The overall completion time of job j is $C_j = C_{t_j j}$. A schedule is feasible if it has the following properties.

(1) Release dates are respected, i.e., $r_j \leq S_{s_j j}$ for each $j \in J$.

(2) Jobs travel towards their destination, i.e., $C_{ij} \leq S_{i+1,j}$ (resp. $C_{ij} \leq S_{i-1,j}$) for rightbound (resp. leftbound) jobs j and $i \in M_j \setminus \{t_j\}$.

(3) Aligned jobs j, j' are not processed on segment $i \in M_j \cap M_{j'}$ concurrently, i.e., $[S_{ij}, S_{ij} + p_j) \cap [S_{ij'}, S_{ij'} + p_{j'}) = \emptyset$.

(4) Opposed jobs j, j' being incompatible on segment $i \in M_j \cap M_{j'}$ are neither processed nor in transit concurrently, i.e., $[S_{ij}, C_{ij}) \cap [S_{ij'}, C_{ij'}) = \emptyset$.

The waiting time of a job j in some schedule is defined as $W_j = C_j - \sum_{i \in M_j}(p_j + \tau_i) - r_j$. Note that optimal solutions wrt. to total (weighted) completion time coincide with optimal solutions wrt. to total (weighted) waiting time since they differ only by a solution independent addend. Hence, NP-hardness results carry over from one to the other. But we have to be careful when it comes to approximation. Nevertheless, the thesis contains results concerning several objective functions.

Our hardness results do not rely on the use of varying speeds and our efficient algorithms assume that the speed of each ship remains equal among all segments. Therefore, we consider within this thesis the following problem.

Bidirectional Scheduling with unit speed (BSP)

Given: A set of n rightbound and leftbound jobs $J = J^r \cup J^l$ with r_j, p_j, w_{js}, s_j, and t_j for each $j \in J$ as well as m segments with $\tau_i, i = 1, \ldots, m$.

Task: Find a feasible schedule minimizing $\sum C_j$, $\sum w_j C_j$, $\sum W_j$, $\sum w_j W_j$, or $C_{\max}$.

1.5 Related work

Decision support for waterway transportation systems has gained already some attention in the literature, mostly on a strategic or tactical base, e.g., on the distribution of investments over time, and often tackled by simulations, see [CB73; ST98; SW05; Mit+13] for a selection. Righini [Rig14] provides a macroscopic view via network flow model on the Northern Italy waterway system to analyze the maximum freight transportation capacity. The author emphasizes that in particular long stretches operated bidirectionally can not be covered by the model. At the operating level that provides concrete decisions for individual ships, there is mainly work concerning the scheduling of lock chambers, see [PT88; VVB09; CS11; Pas+14]. Note that lock scheduling has a bidirectional character, in particular when lock chambers are arranged in sequence. Hence, even though important differences in the concrete models occur, the high level structure exhibits similarities. In our work, we investigate the traffic on the operating level including collision avoidance to gain information for strategic decisions concerning capacity adaptations.

As already mentioned, the STCP itself has many similarities to train timetabling where arrival and departure times of train lines at stations or junctions have to be determined to avoid collisions. We refer to the comprehensive overview on the large body of literature by Lusby et al. [Lus+11]. In general, multicommodity flow formulations in time-expanded networks with additional packing constraints constitute a popular basis for solution approaches. Prior works for special network structures representing long stretches of single and parallel track sections developed models asking for schedules of trains on each track section. This scheduling interpretation for single track networks was first used by Szpigel [Szp73] and spawned certain exact and heuristic methods, mostly based on LP techniques. For a comprehensive summary see [Lus+11, Sec. 2]. Apart from the presence of arbitrary start and target positions on segments in the STCP the most significant difference to train timetabling occurs on the transit segments since the passing of two ships is forbidden for some pairs of ships but not for all.

The train timetable generation often works on aggregated networks and postpones the detailed routing of trains within junctions or stations to a separate planning step. To integrate the STCP's modeling of waiting within sidings we refer to another related application which is the planning of automated guided vehicles (AGVs) described in [Gaw+08]. There, routing requests for single vehicles arrive online over time and quickest dynamic routes without collisions to any of the previously planned vehicles must be computed. A similar collision-free dynamic routing approach can help to route the ships sequentially, one after another. Such a sequential application of dynamic routing optimizes only the arrival time of the currently considered ship and is not able to take care of the global objective. To deal with this drawback, we combine the routing and the scheduling view.

From the theoretical point of view, we provide the new model of bidirectional scheduling that constitutes a generalization of classical machine scheduling. It provides a compact way to represent main components of bidirectional traffic such as for our STCP or bidirectional train traffic on a railway line as described above. To the best of our knowledge, the presented BSP has not been considered in the past nor is it contained as a special case in any other scheduling model. Gafarov et al. [Gaf+15] present a different theoretical model for bidirectional traffic of trains with equal length and speed between two stations. In the model, the connection is split into sub-segments that cannot be used concurrently by any pair of two trains, independent of their directions. In addition, there are distinct relations to other scheduling problems considered in the literature.

Since a set of multiple segments can be considered as sequential machines the BSP is related to non-preemptive shop scheduling problems with machines owning characteristics of both, flow and job shop scheduling. On the one hand, each job has its individual assigned machine subset as in job shop scheduling, on the other hand, the sets are contiguous and ordered as in flow shop scheduling, cf. Lawler et al. [Law+93, Sections 13 and 14] for a survey.

In particular, when all jobs need to be processed on all segments in the same order and all transit times are zero, bidirectional scheduling reduces to flow shop scheduling. Garey et al. [Gar+76] showed that it is NP-hard to minimize the sum of completion times in flow shop scheduling, even when there are only two machines and no release dates. They showed the same result for minimizing the makespan on three machines. Hoogeveen et al. [Hoo+98] showed that there is no PTAS for flow shop scheduling without release dates, unless P = NP. In contrast, Brucker et al. [Bru+05] showed that flow shop problems with unit processing times can be solved efficiently, even when all jobs require a setup on the machines that can be performed by a single server only.

In job shop scheduling, the minimization of the sum of completion times was proven to even be MAX-SNP-hard by Hoogeveen et al. [Hoo+98]. Queyranne and Sviridenko [QS00] gave a $\mathcal{O}((\log(m\mu)/\log\log(m\mu))^2)$-approximation for the weighted case with release dates, where μ denotes the maximum number of operations per job. Fishkin et al. [Fis+03] gave a PTAS for a constant number of machines and operations per job. For fixed m Chakrabarti et al. [Cha+96] provide a randomized $(5.78 + \varepsilon)$- or deterministic $(8 + \varepsilon)$-approximation algorithm that also works in the online setting.

Scheduling jobs bidirectionally in general graphs would be an interesting generalization of our model. Graphs with fixed orientation are considered in packet routing, see the seminal paper by Leighton et al. [Lei+94]. The main focus here lies on bounding the makespan of constructed schedules in dependence of two trivial lower bounds of optimal schedule length, called the congestion and the dilation. Leighton et al. [Lei+94] proved that the makespan of any packet routing problem is linear in congestion and dilation. For more recent progress in this direction, see, e.g., Scheideler [Sch98] and Peis and Wiese [PW11]. Antoniadis et al. [Ant+14] also consider average flow time on a directed line in the online setting with resource augmentation.

Bidirectional scheduling in particular without compatibilities also has similarities to scheduling of two job families with a setup time that is required each time before the processing of jobs of a different family is started. See the comprehensive overviews of Potts and Kovalyov [PK00] and Allahverdi et al. [All+08]. The general comments in [PK00] on dynamic programs for such kinds of problems apply in part to our technique for Theorem 3.3.1. They refer to work of Ghosh [Gho94] for multiple job families being all available at time 0. Ng et al. [Ng+03] consider a single serial batch family with identical processing times and release dates. Batches being processed in parallel are actually related to lock scheduling as pointed out by Coene and Spieksma [CS11]. Therefore, simplified assumptions for a single lock chamber bear dynamic programs of similar character [PT88; CS11]. In addition, Gafarov et al. [Gaf+15] point out related (but not equal) connections of the presented model to scheduling with setup times and develop dynamic programs for several settings, including one for total completion time minimization with release dates.

For a single segment and zero transit times, bidirectional scheduling reduces to the classical single machine scheduling problem without preemption, which Lenstra et al. [Len+77] showed to be NP-hard when minimizing total completion time. For the preemptive variant, scheduling by shortest remaining processing times yields an optimal schedule [Sch68]. In contrast, the weighted total completion time minimization becomes NP-hard for preemptive single machine scheduling [Lab+84]. However, Afrati et al. [Afr+99] showed that both weighted settings admit a polynomial-time approximation scheme (PTAS).

Note that the preemptive SRPT schedule of Schrage [Sch68] is actually an online algorithm that achieves the optimal offline solution. Hence, the best possible competitive ratio here is 1. Hoogeveen and Vestjens [HV96] proved that without preemption no online algorithm can have a competitive ratio better than 2. This value is actually achieved in the unweighted and the weighted case [HV96; Phi+98; AP04]. In the preemptive weighted case, there is a gap between the highest lower bound of 1.073 and smallest upper bound of 1.566 on the best possible competitive ratio [ES03; Sit10a].

Interestingly, despite a considerable effort, the given examples are the only variants of parallel machine scheduling where optimal competitive ratios are known. A whole sequence of papers appeared introducing new algorithms, new relaxations and analytical techniques that decreased the gaps between lower and upper bounds on the optimal competitive ratio [Cha+96; HV96; Goe97; Hal+97; Phi+98; Sei00; Che+01; Goe+02; SS02a; SS02b; SV02; ES03; Lu+03; AP04; MS04; Meg07; CW09; LL09; Chu+10; Sit10a; Sit10b]. We do not intend to give a detailed history of developments; instead, we refer the reader to overviews of Correa and Wagner [CW09] and Megow [Meg07]. Nevertheless, unsatisfactory gaps remain. For the offline setting, polynomial-time approximation schemes have been developed [Afr+99].

To the best of our knowledge, there are only very few problems in online optimization for which an optimal competitive ratio can be determined, bounded, or approximated by computational means. Lund and Reingold [LR94] present a framework for upper-bounding the optimal competitive ratio of randomized algorithms by a linear program. For certain cases, e.g., the 2-server problem in a space of three points, this yields a provably optimal competitive ratio. Ebenlendr et al. [Ebe+09; ES11] study various online and semi-online variants of scheduling preemptive jobs on uniformly related machines to minimize the makespan. In contrast to our model, they assume the jobs to be given one by one (rather than over time). They prove that the optimal competitive ratio can be computed by a linear program for any given set of speeds. In terms of approximating the best possible performance guarantee, the work by Augustine et al. [Aug+08] is closest to ours. They show how to compute a nearly optimal power-down strategy for a processor with a finite number of power states.

2

Solving the
Ship Traffic Control Problem

In this chapter, we present our method to solve the STCP in the most general variant. In addition to the scheduling between sidings we need a feasible allotment of parking slots within sidings over time that is consistent. To that end, we suggest the integration of algorithmic ideas from dynamic collision-free routing of automated guided vehicles. We therefore offer a unified view of routing and scheduling which blends simultaneous (global) and sequential (local) solution approaches to allot scarce network resources to a fleet of vehicles in a collision-free manner. This leads us to construct a fast online heuristic. In view of computational experiments on real traffic data expert planners approved that our combinatorial algorithm is well-suited for the demanded decision support. With the help of instance-dependent lower bounds we assess the quality of our solutions which significantly improves upon manual plans.

This chapter is based on joint work with Marco E. Lübbecke and Rolf H. Möhring [Lüb+14].

In this chapter, we present our method to construct solutions for the STCP in the most general case. The approach reflects the combinatorial and geometric components of the problem as presented in Section 1.3. While feasible itineraries (geometry) imply precedences between sidings (combinatorics), the converse is not true in general. However, for the combinatorial relaxation, we show that itineraries can be constructed from feasible precedences in polynomial time by a dynamic routing algorithm that treats only a single ship at a time (Section 2.1). We extend this quickest path style algorithm to the general situation to insert a single route into an existing plan, respecting all constraints imposed by the geometry, when this is possible (Section 2.2). Yet, this algorithm alone is inappropriate for iteratively constructing solutions by routing ships one by one: we show that any such sequential routing procedure may yield a total waiting time arbitrarily far from optimum (Lemma 2.3.1). The reason lies in the inherent interdependence of the problem's geometry and combinatorics. We thus devise a local search on the precedence decisions (Section 2.3.2) which uses a routing algorithm to incorporate the geometric part. Reflecting the problem's online character we embed the algorithm into a rolling horizon planning (Section 2.3.3).

We evaluate the presented approach with computational experiments on real traffic data considering distinct algorithmic components, solutions attained from historical GPS data and the combinatorial relaxation (Section 2.4). Our solutions improve significantly upon manual plans and are additionally approved by the expert planners on site to be well-suited for the decision support. In addition to the success for our particular application, there are several methodological aspects that are of broader interest:

With our approach we present a possibility to merge algorithmic ideas from the two related applications: the job-shop scheduling interpretation being common in train time-tabling and quickest path algorithms for collision-free routing of automated guided vehicles. Both aspects, the scheduling and the dynamic routing, are interdependently present in our traffic control problem. We show how a dynamic routing algorithm for a single vehicle can be enhanced with scheduling decisions such as to produce a collision-free routing for a whole fleet of vehicles.

Time-space, or *time-expanded* graphs are a standard way to model routing problems over time. In practice, with a fine time discretization, these graphs quickly grow beyond any manageable size. Our application additionally called for a fine *space discretization*, rendering such a modeling approach infeasible. At this point, we emphasize that many approaches in the literature use (time and/or space) discretization already at the modeling stage and thus take a loss in accuracy. This is systematically avoided by our geometric description. Additionally, we have shifted the information complexity from the graph representation to the routing algorithm, resulting in an extremely simple graph, but a more involved algorithm: We propose a Dijkstra-like shortest path algorithm that *implicitly* handles time *and* space discretization with an *arbitrary* precision. It is fair to say that, implicitly, our algorithm chooses, out of a continuum of possible discretizations, exactly *the right* one.

Collision-free routing is a research topic of wide practical interest; besides rail-bound transportation, applications like scheduling industrial robots [RS10; SW11] come to mind. In order to avoid collisions it is customary to block (parts of) trajectories for others for the whole time interval it in use by one robot, even if this is not required in practice. We only forbid *entry times* into these trajectories which enables our implicit space discretization. This allows having several robots on the same trajectory simultaneously and resolves an algorithmic problem which hitherto wasted non-negligible optimization potential.

The geometric and combinatorial problem components naturally suggest a decomposition based approach. While complex geometric feasibility constraints hinder the definition of a *neighbor* in a local search, combinatorial neighbors can be easily computed. We use our dynamic routing algorithm to turn them into improved geometries.

Overall, we solve a very complex practical problem to the fullest satisfaction of the problem owners. Routing over time, scheduling, and (rectangle) packing aspects are to be considered simultaneously, and in an online manner. We developed a decision support tool to evaluate the control of current and future traffic scenarios on the Kiel Canal, respecting numerous constraints at an enormous level of detail. The quality of our planning was verified by experts at different levels. Thus, our tool was used in a study to decide about the enlargement options for the Kiel Canal.

2.1 Realizing a combinatorial frame via iterated routing

For the moment we consider again the combinatorial relaxation as investigated in Section 1.3.2. In our considerations, we have seen that optimal dynamic routes for this relaxation can be constructed by linear programming, once the precedence decisions are fixed. However, as we show next, this *global* view, i.e., *simultaneously* deciding about all visit times is not necessary, but rather *local* decisions for a single ship at a time suffice to create optimal collision-free dynamic routes for our combinatorial relaxation.

Assume we are already given dynamic routes $\mathcal{P}^*$ for a subset $S^* \subseteq S$ of ships, and we would like to construct a dynamic route for another ship $i \notin S^*$, respecting $\mathcal{P}^*$. This can be done in a *greedy* manner. Define the visit times of i consecutively for each segment as early as possible after its arrival without violating vertical distances to dynamic routes in $\mathcal{P}^*$. This yields a collision-free dynamic route with minimum waiting time for ship i w.r.t. $\mathcal{P}^*$ and takes polynomial time. When a conflict occurs on a transit segment E, traversing E as early as possible implies waiting at the end of a siding, directly in front of E. This is feasible as constraints regarding parking and vertical distances are relaxed in sidings.

We focus on solutions with a certain structure motivated by this greedy routing. Therefore, we call a dynamic route P_i of a considered solution $\mathcal{P}$ to be *earliest arrival* w.r.t. the other dynamic routes of $\mathcal{P} \setminus \{P_i\}$, if the arrival time $\underline{\tau}_i(p)$ of the dynamic route P_i is at each position $p \in [s_i, t_i]$ as early as possible with respect to $\mathcal{P} \setminus \{P_i\}$ preserving collision-freeness and the solution's combinatorics. Hence, the corresponding ship waits at no position longer than necessary, and all waiting of this ship takes place as late as possible. Note that the earliest arrival route of a ship with respect to the others and their combinatorics is unique. A solution $\mathcal{P}$ is called *earliest arrival* if each dynamic route of $\mathcal{P}$ is earliest arrival w.r.t. the other dynamic routes of $\mathcal{P}$. A solution can be turned into an earliest arrival solution by iteratively turning dynamic routes into earliest arrival ones in an arbitrary order (that may contain cycles) without increasing the waiting time of any ship. Observe that greedy routing produces an earliest arrival route for a single ship with respect to the already planned ones. Furthermore, greedy routing can be extended to respect additionally given predecessors of the current ship on each segment by defining the visit times of each segment as early as possible without violating them. We use this fact to produce earliest arrival solutions only given the combinatorics.

Theorem 2.1.1. *For our combinatorial relaxation, there is a polynomial time algorithm using greedy routing that, given an instance (C, S, R) and a combinatorial frame ζ on S, either creates a feasible earliest arrival solution $\mathcal{P}$ realizing ζ or proves that no realization exists.*

We constructively prove the theorem by describing Algorithm REALIZE. In Section 2.3.2 we generalize it to respect siding constraints. Define a *partial dynamic route* of ship $i \in S$ to be a dynamic route of i with the relaxation that the first and the last x-coordinates can be any points in $[s_i, t_i]$, not necessarily on the boundary. A partial dynamic route of ship $i \in S$ is called a *prefix route* if the first x-coordinate equals s_i and a *suffix route* if the last x-coordinate equals t_i. Consider some $(j, i) \in \zeta(E)$ for a transit segment E, i.e., ship j has to precede ship i on E. To ensure this precedence the visit time of j on E must be known before deciding about the visit time of i on E via greedy routing. This can be achieved by producing partial dynamic routes for each ship in distinct steps, that fit together, instead of one complete dynamic route in one step. This is done in Line 4 of Algorithm REALIZE which works as follows:

(1) Let p_i be the last x- and τ_i be the last y-coordinate of the current partial dynamic route P_i of i.

(2) Find a collision-free suffix route P_i' for i with start position p_i and start time τ_i that respects the current partial routes $\{P_j \mid j \in S\}$ and the given combinatorics ζ.

(3) Let E be the first transit segment on P_i' such that $(j, i) \in \zeta(E)$ and E is not yet contained in the current prefix route P_j. Denote the beginning of the siding before E on P_i by p'.

(4) Extend the prefix route P_i by P_i' until the position p'.

The location of p' is well-defined only if it yields a proper extension, i.e., if there are at least two sidings between the end of P_i and E. We say in this case that ship i is *free*. This can be tested (or also administrated together with the route extensions of the other ships) in polynomial time.

---

input: canal C, requests R for ships S, combinatorics ζ
output: an earliest arrival solution $\mathcal{P}$ realizing ζ or the information that no
         realization of ζ exists
1   for each $i \in S$ initialize a partial dynamic route $P_i := ((s_i, r_i))$
2   **while** there are ships S' with incomplete dynamic route:
3   **if** there is a free ship $i \in S'$:
4   extend P_i by greedy routing of i as far as possible with respect to ζ
5   **else**:
6   **return** "no realization for ζ"
7   **return** $\{P_i \mid i \in S\}$

Algorithm REALIZE: Realizing the given combinatorics in our combinatorial relaxation.

It remains to be shown that Algorithm REALIZE is correct according to the requirements of Theorem 2.1.1.

Lemma 2.1.2. *Algorithm REALIZE produces an earliest arrival solution realizing ζ iff ζ is realizable for the instance (C, S, R).*

Proof. If Algorithm REALIZE returns a solution it is collision-free, earliest arrival, and realizes ζ by construction. Consequently, ζ is realizable.

For the converse direction assume that there is a realization $\mathcal{P}$ of ζ. W.l.o.g. $\mathcal{P}$ is an earliest arrival solution. We prove the following loop invariant for Algorithm REALIZE:

Claim. As long as S' is not empty there is at least one free ship $i \in S'$ in each iteration and the constructed extension of P_i is equal to the corresponding part of i-th dynamic route in $\mathcal{P}$.

This implies that the algorithm terminates with a complete collision-free dynamic route for each ship which finally equals the one of $\mathcal{P}$.

The proof is by induction over the number of iterations. The base case is obvious. For the inductive step assume that the claim holds for all previous iterations and $S' \neq \emptyset$. By contradiction, suppose that there was no free ship, i.e., each ship i with incomplete route has still an open predecessor j w.r.t. ζ on the next transit segment of its request after the current end of P_i. When each ship has a predecessor blocking its route extension there must be a cycle in ζ since S' is finite. However, since ζ is realizable it is thus necessarily acyclic, a contradiction. Thus, there must be a free ship i in this iteration. By the inductive hypothesis, all prefix routes constructed in earlier iterations are equal to the corresponding parts in $\mathcal{P}$. Hence, all prefix routes of the other ships preceding the current free ship on the transit segments of its next routing part within $\mathcal{P}$ are already given and prevent ship i from traversing them too early. Hence, the corresponding unique earliest arrival route part constructed by the algorithm equals the part of the corresponding earliest arrival route of $\mathcal{P}$. □

For a realizable combinatorial frame of a given instance the output of Algorithm RE-ALIZE is well-defined; we thus conclude from the above induction hypothesis:

Corollary 2.1.3. *For any feasible solution to our combinatorial relaxation there exists a unique earliest arrival solution which can be constructed by Algorithm REALIZE.*

The local view on routing that we have adopted in this subsection will greatly help us in the following to incorporate feasibility constraints concerning passing and waiting in sidings when dealing with the original, unrelaxed problem.

2.2 Collision-free routing for a single ship

Assume that we are given collision-free dynamic routes $\mathcal{P}'$ for a subset of ships $S' \subseteq S$. We would like to construct a dynamic route for *a further ship* $i \in S \setminus S'$ which is collision-free and minimizes its waiting time w.r.t. the already given $\mathcal{P}'$. With siding conditions relaxed (Section 2.1) this insertion was easy via greedy routing. In the original setting, more work is needed for two reasons: First, if a conflict occurs, the siding directly in front of it may be fully occupied such that the waiting must take place at an earlier siding. It is thus not always possible to use the earliest available time to traverse a transit segment. Second, the routing algorithm must now take spatial decisions concerning collision-free waiting within sidings.

Our setting is related to the iterative routing of automated guided vehicles (AGVs) where transportation requests arise over time in a graph. The routing algorithm of Gawrilow et al. [Gaw+08] answers the respective next request, avoiding collisions to previously routed AGVs already during the route computation. Confer also the related dissertation of Stenzel [Ste08]. Despite significant differences concerning the practical contexts, we extend their algorithmic idea to plan ship i optimally with respect to the dynamic routes of $\mathcal{P}'$. It is worth noting that shortest path problems with time windows are widely studied as subproblem in the vehicle routing literature, see [Des+95; SD88] for overviews. The permanent labeling algorithm developed by Desrochers and Soumis [DS88] (see also [Des+95]) is the basis for the algorithm of [Gaw+08] which runs in polynomial time for the objective of finding a *quickest* path respecting time windows.

Like in Dijkstra's algorithm [Dij59] we maintain labels representing arrival times of the ship, cf. Algorithm DIJKSTRA of Section 1.1 or [Cor+09]. However, instead of referring to a single point in time, our labels correspond to feasible arrival time *intervals*. Further, for each arc, we store *forbidden time windows* during which it is not allowed for the ship to enter that arc in order to avoid collisions with conflicting ships. Together, this gives us an *implicit* time-expansion of the graph and avoids the rather traditional way of guaranteeing collision-freeness by deleting arcs from a time-expanded graph, confer Section 1.1. When exploring the outgoing arcs of a node the label intervals will be adapted according to the forbidden time windows. The collection of labels then covers all possibilities to traverse a transit segment, not only the earliest one. This resolves the first algorithmic issue stated in the introduction of this section.

Algorithm CFR gives an overview of our quickest path computation with time windows. A *label lab* $= (a, [epat, lpat], pred, \ell)$ represents a *set* of collision-free dynamic routes for ship i, from its start position s_i to the target node of arc a, with an interval of earliest ($epat$) and latest ($lpat$) possible arrival times, and a reference to a predecessor label *pred* of *lab* on all represented routes. Using *pred*, a route with arbitrary arrival time in $[epat, lpat]$ can be reconstructed recursively. Since the length ℓ of some arcs will be adapted dynamically by the algorithm, each label holds ℓ as a further component. It is possible that a label interval degenerates to a single point in time. The precise construction of the graph $G(C)$ and the maintenance of forbidden time windows $\mathcal{F}(.)$ is described later.

---

input: canal graph $G(C)$, routing request $r_i \in R$, time windows $\mathcal{F}(.)$
output: quickest path P_i w.r.t. $\mathcal{F}(.)$

1   enqueue start-labels for s_i in priority queue PQ
2   **while** PQ is not empty:
3       $cur :=$ best label w.r.t. $\prec$ dequeued from PQ
4   **if** cur corresponds to t_i:
5   **return** reconstructed dyn. route for cur with earliest possible arrival time
6   **foreach** successor arc $succ$ of label cur:
7   **foreach** label $next$ resulting from propagation of cur along $succ$ w.r.t. $\mathcal{F}(.)$:
8   **if** $next$ is not dominated by an active label of $succ$:
9   enqueue $next$ in PQ
10   remember $next$ as an active label of $succ$
11   delete active labels of $succ$ that are dominated by $next$ also from PQ

Algorithm CFR: Calculate a quickest path for ship i avoiding given forbidden time windows $\mathcal{F}(.)$ within a canal graph $G(C)$.

Like in Dijkstra's algorithm, the union of all labels is organized in a priority queue. We use a natural order $\prec$ on the labels which is induced by the earliest possible arrival times $epat$. For labels which are identical according to this order, preference is given to certain properties, see below. Controlled by this earliest-label-first order, the arcs are explored starting from labels corresponding to the arc that contains the requested start position until an arc containing the target position is reached. Two important sub-tasks distinguish the algorithm from Dijkstra's classic (details follow):

(1) The forbidden time windows are respected during the exploration of an arc resulting in a set of split labels. Furthermore, for arcs corresponding to a waiting track, the label interval is expanded to represent the waiting. Together, this is called *propagation*.

(2) Comparing two labels corresponding to one arc, we cannot always determine which one will produce no better solutions than the other. In this case, both must be remembered for further consideration. Otherwise, a *dominated* label can be deleted.

In the remainder of this section we specify the details to apply the algorithm in our canal setting: define an appropriate graph, develop time windows guaranteeing collision-freeness, and describe the distinct steps of the algorithm. In particular, this answers the second algorithmic challenge concerning waiting decisions within sidings.

2.2.1 Graph for collision-free routing

The quickest path computation must take place in an appropriate graph. Our use of forbidden time windows in the algorithm saves us from discretizing the time-horizon and from forming a time-expanded graph. However, another complication arises from the spatial dimension when modeling sidings and their waiting tracks. Recall that each ship traverses the siding on the passage track, changes to the waiting track if it reaches the defined waiting position, and later continues its passage again on the passage track. It seems natural to partition the waiting track into a discrete set of "parking lots," or waiting arcs, to and from which a ship can branch from arcs corresponding to the passage track. Unfortunately, this would cause several problems. With a fixed "reserved space" short ships would consume more waiting capacity than necessary. This would result in a loss of siding capacity. Answering to this by partitioning the waiting track in many short waiting arcs would also increase the number of labels and hence, the running time. Worse, a parking ship would then use several arcs, but the propagation along an arc should work locally, i.e., the reduction of the label interval to ensure collision-freeness should only depend on the time windows saved for that particular arc. For this, we would need to determine, after a route computation, which other arcs were affected when a single waiting arc is used by a ship for waiting and hence, on which arcs to store the resulting forbidden time interval. This could be handled when we assumed identical ship dimensions as is the case for AGVs in [Gaw+08]. In our setting, however, many individual ship lengths vary from 50 to 250 meters. Finally, no partition of the waiting track could not ensure that a ship waits at most once per siding. For these reasons, we refrain from statically partitioning the waiting tracks, i.e., we do not use a spatial discretization at all. Instead, we let our algorithm dynamically consider waiting positions as necessary.

Figure 2.1: Graph $G(C)$ representing canal C. In a siding, different types of arcs reflect time/space before (type α), while (type β), and after (type γ) waiting.

Figure 2.1 displays how we translate a canal into a graph. A node is introduced on the boundary points of each segment. For each segment, these two nodes are connected by a right- and a leftbound arc, respectively. In order to model the possibility to wait on the waiting track, we introduce a further node per direction within each siding. It allows the ship to interrupt the passage of the siding for waiting and continue the passage later on. The waiting track itself is represented by a loop incident to this node. Hence, there are three arcs per siding and direction, one of *type* α for the passage before waiting, one of *type* β for waiting on the waiting track, and one of *type* γ to continue the passage after waiting. In contrast to the nodes at the segment boundaries the node representing waiting in a siding has no explicit embedding: The waiting arc does not correspond to a fixed waiting position but will hold information about all ships of $\mathcal{P}'$ waiting within that siding. As part of the propagation process along arc α all necessary but not more positions for the waiting node will be checked depending on this information. Given a canal C, we denote this graph as $G(C)$. All the potential complexity of this graph—we neither have time nor space discretization—is avoided by entirely shifting the information management to the algorithm.

2.2.2 Forbidden time windows

Consider a transit segment $E \in \mathcal{E} \setminus \mathcal{T}$. As suggested by Equation (1.12) the dynamic route of ship $j \in S'$ (in conflict with the new ship i on E) defines an upper bound $u_j := d_{jE} - \Delta(i, j, E)$ on the visit time of i on E before the passage of ship j. Similarly, a lower bound $\ell_j := d_{jE} + \Delta(j, i, E)$ for visiting E after j arises. This results in a *time window* $F_j := [u_j, \ell_j]$ implied by the dynamic route of j during which it is *forbidden* for ship i to enter E, i.e., we have to ensure that $d_{iE} \notin F_j$. As before, the interval is with respect to the start position of segment E to have a reference point independent of s_i. In this sense, we can determine for each arc and each dynamic route in $\mathcal{P}'$ time windows that restrict the use of this arc to ensure a collision-free dynamic route for i.

Now consider a siding $[x, y] =: T \in \mathcal{T}$. A ship that waits in T does not block the passage track of T. Therefore the passage of that track before waiting and the passage after waiting must be considered separately. Consider a ship $j \in S'$ with given dynamic route that waits at position $p_j \in T$ and has a conflict with i. If ship i's waiting position p_i is, say, in front of p_j we must ensure that the passage of ship i before its waiting does not collide with the passage of j before its waiting and furthermore with the passage of j after its waiting. Hence, depending on the considered waiting position p_i ship j can induce two forbidden time windows for the corresponding arc of type α or two for the corresponding arc of type γ. For more details confer Figure 2.2. Additionally, the waiting ship j produces a forbidden time window on the waiting arc of T (type β) for the interval $\tau_j(p_j)$ if the waiting positions p_j and p_i have a distance smaller than $h_i + h_j$.

To summarize, the set of dynamic routes $\mathcal{P}'$ yields respective sets of forbidden time windows $\mathcal{F}(a, i, p_i)$ for siding arcs a, and respective sets of forbidden time windows $\mathcal{F}(a, i)$ for transit arcs a. These restrict the use of a for ship i as described, possibly depending on the given waiting position p_i.

As already mentioned, we want to have fast local access to the set of time windows of an arc a during the propagation along a. To this end, each arc holds for each given dynamic route $P_j \in \mathcal{P}'$ the respective local information to define a time window (the visit time of the segment, possibly the waiting position within the siding, and a link to the properties of the corresponding ship j). Since each time window additionally depends on properties of ship i it can be calculated in constant time.

2.2.3 Routing details for the canal

We now explain the single steps of Algorithm CFR in more detail.

Propagation. The core task of propagating a label *cur* along an arc *succ* with respect to $\mathcal{F}(.)$ (Line 7 of Algorithm CFR) is the splitting of a label interval corresponding to the

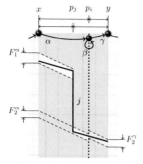

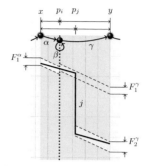

(a) The tested waiting position p_i is behind the fixed waiting position p_j.

(b) The tested waiting position p_i is in front of the fixed waiting position p_j.

Figure 2.2: A given dynamic route of ship j defines a set of time windows in sidings. If ship i is supposed to wait behind ship j (part (a)) there are two time windows where i is not allowed to enter T. Otherwise (part (b)) ship j induces two time intervals where ship i is not allowed to continue its passage after waiting. To determine the boundaries of the forbidden time windows we make use of $\Delta(\cdot)$ as above. Since it is possible that ship i and ship j have different velocities, we must restrict the domains of i and j valid for the calculation of $\Delta(\cdot)$ for the distinct cases (indicated by the light areas): F_1^α corresponds to $[x, p_i]$ and $[x, p_j]$, F_2^α to $[x, p_i]$ and $[p_j, y]$, F_1^γ to $[p_i, y]$ and $[x, p_j]$, and finally F_2^γ to $[p_i, y]$ and $[p_j, y]$ if the respective intervals are not disjoint (cf. Figure 1.8). If the start- or target position of i or j is situated within this siding, the request intervals must be further reduced appropriately. For reasons of clarity we located the time windows of the arcs of type γ in this picture w.r.t. the end of T. Note that we actually define them translated to the start of T such that they are compatible with all other definitions.

given forbidden time windows into a set of new labels for arc $succ$ guaranteeing collision-free dynamic routes. This splitting is described in Algorithm SPLIT the input of which must be chosen according to the type of the considered arc $succ$.

---

input: arc $succ$, label cur, time value $newend$, union of time intervals F, label length ℓ, transit time τ
output: Set of feasible labels for $succ$ of length ℓ with cur as predecessor
1 $\mathcal{I} := [epat(cur), newend] \setminus F$ // open time windows
2 $\mathcal{L} := \emptyset$
3 **foreach** maximal interval $[epat', lpat']$ of $\mathcal{I}$ with $epat' \le lpat(cur)$:
4 $\quad \mathcal{L} := \mathcal{L} \cup (succ, [epat' + \tau, lpat' + \tau], cur, \ell)$
5 **return** $\mathcal{L}$

Algorithm SPLIT: Splitting a predecessor label along an arc $succ$ into a set of feasible labels for $succ$.

If $succ$ corresponds to a transit segment E, the algorithm is called with the following arguments: $newend := lpat(cur)$, F as union of $\mathcal{F}(succ, i)$, label length $|E|$, and transit time τ_{iE}. Since F contains each point in time at which entering of E is prohibited for ship i, each time value of $\mathcal{I}$ is a feasible entering time for i. Since the new labels correspond to the target node of $succ$, the time intervals must be translated to this position.

Is $succ$ of type α for a siding T, the splitting must be done several times, namely for each *interesting* waiting position p_i. Each iteration calls Algorithm SPLIT with $newend := lpat(cur)$, $F :=$ union of $\mathcal{F}(succ, i, p_i)$, label length $:= p_i$, and the resulting transit time for this length as input. To determine the set of all interesting waiting positions, each maximal time interval with invariant waiting conditions within T from the earliest arrival of cur is considered. In this, each free waiting area will be filled with waiting positions respecting the horizontal parking distances with h_i as step size from the end of the siding to its beginning until one waiting position yields a label who's interval would not be reduced by the splitting process on any of the three siding arcs. All further waiting positions would not yield further possible arrival times at the end of the siding and hence, would not yield a better solution. This reduces the number of constructed labels often drastically and is important for an acceptable running time.

The input value $newend$ becomes important, if $succ$ is an arc of type β for siding T. In this case, the waiting will be modeled by an extension of the interval to $newend := \infty$. After this extension it will be split by F as union of $\mathcal{F}(succ, i, \ell(cur))$ since $\ell(cur)$ is the tested waiting position. To provide this information also for the next arc, the label length is also set to $\ell(cur)$ (and not 0). But the transit time is 0.

Finally, we describe the propagation for the case that $succ$ corresponds to an arc of type γ for siding T. In this case, label cur can correspond to arcs of both types α and β. The start node of $succ$ then has the corresponding position and hence, the given label length must be $|T| - \ell(cur)$ and the transit time, respectively. Furthermore, each time window of $\mathcal{F}(a, i, \ell(cur))$ must be translated by $\ell(cur)$ w.r.t. the velocity of i for the construction of F. The value $newend$ is again set to $lpat(cur)$.

Dominance. A label lab is said to *dominate* another label lab' if both represent the same position in the canal, and lab' cannot lead to any better routes than lab. Dominated labels can be deleted. More precisely, label lab dominates lab' if and only if both correspond to the same arc a, both correspond to the same length ℓ if a is of type α or β, and

$$lab \preceq lab' \quad \text{and} \quad [epat(lab'), lpat(lab')] \subseteq [epat(lab), lpat(lab)].$$

Testing the dominance of labels (Lines 8 and 11 of Algorithm CFR) helps to reduce the number of labels that must be propagated without losing any dynamic route of minimum waiting time. Confer Theorem 2.2.3 of Section 2.2.4.

Resulting Dynamic Route. The first constructed label that corresponds to an arc containing the target position of ship i (Line 5 of Algorithm CFR) represents a *set* of dynamic routes with smallest possible arrival time and hence, with minimum waiting time. It may still be open in which of the sidings in front of an occurred conflict to spend the waiting time. Each route in the set can be constructed recursively from the last label in reverse order. Hence, we have the freedom to choose among these optimal dynamic routes and we decide to let waiting always take place within the latest possible siding. Also, the ordering $\prec$ of the priority queue is defined such that late waiting positions within each siding are preferred. Consequently, in the case of relaxed siding conditions, Algorithm CFR produces the same earliest arrival solutions as greedy routing.

Note that *earliest arrival* routes are not well-defined anymore. If a waiting-position in the front of a siding is occupied for a part of the necessary waiting interval it is either possible to wait behind this position or to wait in an earlier siding until waiting in the front position is possible. In both cases, the earliest arrival property is violated at some position.

After the reconstruction of the dynamic route of i we update the corresponding information on each arc for future calculations of time windows.

When there is not enough capacity available for waiting within the sidings, the priority queue runs out of labels before a label corresponding to the target position is reached. In that case we conclude that a dynamic route for ship i does not exist.

Knapsack Relaxation. Assuming relaxed siding conditions, a solution can be realized from the given combinatorics via greedy routing (Theorem 2.1.1). When we additionally impose the siding capacity as a knapsack constraint, it might be impossible for each ship to wait directly in front of a conflict. Thus, greedy routing may produce infeasible solutions. Even so, Theorem 2.1.1 can be extended to this stronger knapsack relaxation (which is relevant in the case of all identical ship dimensions) when using collision-free routing instead of greedy routing: only time windows and propagation on arcs of type β need to be slightly adapted.

Corollary 2.2.1. *For the knapsack relaxation, there is a polynomial time algorithm using collision-free routing that, given an instance (C, S, R) and a combinatorial frame ζ on S, either creates a feasible earliest arrival solution $\mathcal{P}$ realizing ζ or proves that no realization exists.*

2.2.4 Running time

Determining the running time of the routing procedure includes to take care of the number of waiting positions that need to be tested. This number is strongly related to the maximum ratio ρ_{h_i} of siding length and horizontal distance of ship i. With some extension of the argumentation in [Gaw+08], it can be shown that the running time is polynomial in the maximum number of time windows η per arc, the number of segments, and ρ_{h_i}. Note that η is bounded by $2n$, and often much smaller. For the real world instances coming from the Kiel Canal we can assume that each ratio ρ_{h_i} is bounded by a constant. For bounding the running time observe that any label that is dominated at some point is not dequeued from the priority queue for further propagation.

Lemma 2.2.2. *The number of labels that are never dominated within Algorithm CFR is bounded for each arc by $m^2 \eta^3 \rho_{h_i}^2$.*

Proof. We start with some general observations that are needed to finally count the non-dominated labels per arc. For the analysis we consider how labels propagate along certain paths with fixed arc length within the graph where each arc has fixed length. Therefore, we define in this setting a path $P = ((a_1, \ell_1), \ldots, (a_p, \ell_p))$ to be a sequence of incident arcs together with corresponding length values (with an analog interpretation as for the labels). The set of produced successor labels of some label *cur* on an arc a is denoted by $Succ(cur, a)$ or $Succ_P(cur, a)$ if we consider only those successor labels of cur produced on a certain path P. Denote by $Dom(cur, a) \subset Succ(cur, a)$ and $Dom_P(cur, a) \subset Succ_P(cur, a)$ the respective subsets of those labels that are dominated at some point. To get the union of successors for a set of labels, a set of paths or a set of arcs we simply plug in the corresponding set to the respective argument, slightly abusing notation.

Observation 2.1. We first analyze how a set of disjoint labels propagate on a simple path of arcs with fixed length without waiting. To be more precise, we consider some path $P = ((a_1, \ell_1), \ldots, (a_p, \ell_p))$ without an arc of type β and a set of labels $\mathcal{L}$ for $(a_j, \ell_j) \in P$ with disjoint label intervals. Since the length of no interval is increased by the propagation and they are all translated by the same transit time, all intervals of the labels in $Succ_P(\mathcal{L}, a_k)$ are disjoint for each arc $a_k, k > j$ on P. Hence, no time window can split more than one label interval into two and the number of successor labels $Succ_P(\mathcal{L}, a_k)$ on each $a_k, k > j$ is bounded by $\#\mathcal{L} + (k - j)\eta$.

We now consider a siding $T \in \mathcal{T}$ between start and target position of ship i. The corresponding arcs on T of each type in heading of i are denoted as $a^\alpha(T)$, $a^\beta(T)$, and $a^\gamma(T)$. Let $\mathcal{W}(T)$ be the set of waiting-positions that are tested by Algorithm CFR. Let furthermore $\mathcal{L}(T)$ be the set of all labels arriving for propagation through T, i.e., all non-dominated labels on the incoming arc before $a^\alpha(T)$.

Observation 2.2. We start by bounding the number of tested waiting positions $\mathcal{W}(T)$. Since we have at most η time windows on $a^\beta(T)$ we can bound the number of maximal time intervals with invariant waiting conditions within T by $(2\eta + 1)$. Therefore, there are at most $\rho_{h_i}(2\eta + 1)$ possible waiting positions in T.

Observation 2.3. We investigate the labels that are produced when a label $cur \in \mathcal{L}(T)$ is propagated through T without waiting, i.e., without using arc $a^\beta(T)$. Let $\mathcal{P}_2(T)$ be the set of paths $P_i := ((a^\alpha(T), p_i), (a^\gamma(T), |T| - p_i))$ with only the two passage arcs defined by the tested waiting positions $p_i \in \mathcal{W}(T)$. For geometrical reasons (cf. Figure 2.2) we can observe that the set of successor labels $Succ_{P_i}(cur, a^\gamma(T))$ on arc $a^\gamma(T)$ corresponds for each $P_i \in \mathcal{P}_2(T)$ to the same set of label-intervals. Since all but one label per interval are dominated, we get a resulting set of non-dominated labels $Succ_{\mathcal{P}_2(T)}(cur, a^\gamma(T)) \setminus Dom_P(cur, a^\gamma(T))$ with a cardinality equal to each $\#Succ_{P_i}(cur, a^\gamma(T)), P_i \in \mathcal{P}_2(T)$.

Observation 2.4. Finally, we analyze all labels that are produced on arc $a^\beta(T)$, i.e., the waiting track of T. Let $\mathcal{P}_3(T)$ be the set paths containing for each tested waiting positions $p_i \in \mathcal{W}(T)$ the path $P_i := ((a^\alpha(T), p_i), (a^\beta(T), p_i), (a^\gamma(T), |T| - p_i))$ with all three siding arcs. We consider some path $P_i \in \mathcal{P}_3(T)$ for waiting position $p_i \in \mathcal{W}(T)$ and let $\mathcal{I}$ be the set of all maximal time intervals not blocked by the forbidden time windows of $\mathcal{F}(a^\beta(T), i, p_i)$. Note that $\mathcal{I}$ contains at most $\eta + 1$ intervals and that they are disjoint. For each label $lab \in Succ_P(\mathcal{L}(T), a^\beta(T))$ there is a unique interval $I \in \mathcal{I}$ such that the label interval of lab is contained in I. Let $\mathcal{L}(I), I \in \mathcal{I}$ be the corresponding partition of the labels in $Succ_P(\mathcal{L}(T), a^\beta(T))$ by their respective intervals. We consider some interval $I \in \mathcal{I}$ and observe that $lpat(lab)$ is equal to the right bound of I for each $lab \in \mathcal{L}(I)$. Therefore, there is a label lab_I such that $[epat(lab), lpat(lab)] \subseteq [epat(lab_I), lpat(lab_I)]$ and $lab_I \preceq lab$ for each $lab \in \mathcal{L}(I)$. Hence, all but one label per interval are dominated which yields a set $\mathcal{L}_{p_i}^\beta(T) := Succ_{P_i}(\mathcal{L}(T), a^\beta(T)) \setminus Dom_{P_i}(\mathcal{L}(T), a^\beta(T))$ of at most $\eta + 1$ non-dominated labels on arc $a^\beta(T)$ with disjoint label intervals. The union $\mathcal{L}^\beta(T)$ over all $P_i \in \mathcal{P}_3(T)$ then has by Observation 2.2 a cardinality of $O(\eta^2 \rho_{h_i})$. In particular, this upper bound is independent from the number of arriving labels $\mathcal{L}(T)$.

We now combine the single observations to bound the overall number of labels per arc. To this end, we give a decomposition of the set of paths corresponding to the developed labels, cf. Figure 2.3. Each path of this decomposition contains either no arc of type β or exactly one arc of type β which is then the first arc of this path. We use that we can control what is happening on such a path due to our observations.

Figure 2.3: Illustration of a path decomposition such that each created label can be located on one of these paths.

We first consider the set of all paths $\mathcal{P}$ from s_i to t_i using no arc of type β for waiting. Note that only constantly many start labels are enqueued. Consider a start label $slab$ and a path $P \in \mathcal{P}$. By Observation 2.1 each arc a on P gets $O(m\eta)$ many successor labels $Succ_P(slab, a)$. By Observation 2.3 about dominance on arcs of type γ, each considered waiting position per siding creates the same set of label intervals and only one label per interval is not dominated and produce successors. This yields for the union of paths $\mathcal{P}$ that the number of labels in $Succ_P(slab, a)$ remains in $O(m\eta)$ for each arc a corresponding to a transit segment and is by Observation 2.2 in $O(m\eta^2\rho_{h_i})$ for each arc a of type α or γ.

We now consider for each siding T between s_i and t_i and each waiting position $p_i \in \mathcal{W}(T)$ the set of all paths $\mathcal{P}_{p_i}(T)$ from $(a^\beta(T), p_i)$ to the last arc $a(t_i)$ corresponding to t_i using no further arc of type β for waiting. Note that by Observation 2.4 each set $\mathcal{L}_{p_i}^\beta(T)$ contains at most $\eta + 1$ non-dominated labels with disjoint intervals that do produce successor labels. With an analog reasoning as above we get by Observations 2.1, 2.2, and 2.3 that the cardinality of each $Succ_{\mathcal{P}_{p_i}(T)}(\mathcal{L}_{p_i}^\beta(T), a)$ is in $O(m\eta^2\rho_{h_i})$.

It remains to observe that for each label lab there is either a start label $slab$ such that $lab \in Succ_P(slab, a(lab))$ or a siding T and a waiting position $p_i \in \mathcal{W}(T)$ such that $lab \in Succ_{\mathcal{P}_{p_i}(T)}(\mathcal{L}_{p_i}^\beta(T), a(lab))$. Since we have less than m sidings and $O(\rho_{h_i}\eta)$ many waiting positions per siding the combination yields $O(m^2\eta^3\rho_{h_i}^2)$ many non-dominated labels per arc. □

Theorem 2.2.3. *Algorithm CFR finds a collision-free dynamic route for ship i with minimum waiting time wrt. the dynamic routes of $S' \subset S$ in polynomial time.*

Proof. First, consider the chain of succeeding labels arriving at a target label. For each two successive sidings T_1, T_2 we are able by construction to connect each two label intervals of $a^\beta(T_1)$ and $a^\beta(T_2)$ by a straight line with slope corresponding to i-th velocity that is crossing each intermediate label interval at the corresponding position. The same holds for the start label and the first siding and the last siding and the target label where in particular each moment of the label interval can be connected (or simply between start and target label). Furthermore, the corresponding time-values are non-decreasing wrt. the heading of i. Hence, we can construct corresponding dynamic routes. These dynamic routes with corresponding visit times within label intervals are guaranteed by the forbidden time windows to respect all vertical and horizontal distances to all dynamic routes of the other ships of S'. Hence, they are collision-free. Let C_i^* be the earliest point in time ship i can arrive its target position without collisions wrt. the given dynamic routes for S'. Then, there is for each segment $E \in \mathcal{E}$ and each moment $t \leq C_i^*$ where i can arrive without collisions at the end of E a corresponding label containing t since no such point in time is dominated. Consequently, the earliest created target label yields a dynamic route with minimum waiting time.

Finally, we consider the running time. The number of labels per arc that are not dominated also bounds the number of concurrently active labels per arc to $O(m^2\eta^3\rho_{h_i}^2)$ by Lemma 2.2.2. Hence, the number of labels concurrently contained in the priority queue as well as the total number of labels that can be dequeued is bounded by $O(m^3\eta^3\rho_{h_i}^2)$. Therefore, the algorithm has $O(m^3\eta^3\rho_{h_i}^2)$ many iterations. Each iteration is then polynomial in $O(\rho_{h_i}\eta)$ for the propagated labels times $O(m^3\eta^3\rho_{h_i}^2)$ for the dominance tests and updates. □

2.3 A heuristic for the STCP

2.3.1 Construction of solutions by sequential routing

We use a natural idea [Gaw+08] to construct an initial solution from scratch: We iteratively route ships through the canal, one after another, using collision-free routing. We sort ships by non-decreasing times they (would have) entered the canal (i.e., the translation of each release date from the start position to the canal boundary, if needed). This is a canonical ordering to route ships one by one, but it is somewhat arbitrary, and there may be better orderings which give better overall solutions. However, the question for a *best* ordering of such a *sequential routing* is void in view of the following lemma.

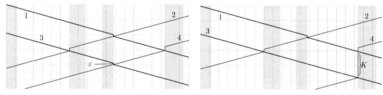

(a) Each ship is waiting for a small amount. (b) One ship without waiting induces high waiting time for another.

Figure 2.4: Example where sequential routing yields high waiting time independent of the chosen routing sequence.

Lemma 2.3.1. *Any sequential routing procedure can yield arbitrarily bad solutions, even if all horizontal distances between ships are zero.*

Proof. Figure 2.4 schematically sketches an instance where in each optimal solution *each* ship has to wait for a short time of, say, at most ε. Such a solution contains a cyclic waiting pattern which cannot be achieved via any sequential routing procedure: at least the ship which is routed first will not wait anywhere. However, such a ship causes another ship to wait for at least $K \gg \varepsilon$. With $\varepsilon \to 0$, one cannot avoid an arbitrarily large deviation from the optimum when routing sequentially. □

2.3.2 Improving schedules by local search on the combinatorics

We have seen already earlier that decisions about who is waiting for whom and where constitute the core difficulty of our problem. The most important insight from Lemma 2.3.1 is that, regardless of the ordering of ships, sequential routing is limited in the structure of schedules that it is able to produce. Actually, it does not actively produce a schedule at all, i.e., the combinatorics of the solution comes as a necessary by-product. We therefore need to find ways to produce different, in a sense *richer* combinatorial frames, and we need to be able to deal with this in conflict-free routing. We address the second item first.

Algorithm REALIZEb. A given combinatorial frame may be impossible to realize because of the siding capacities. The following modification of Algorithm REALIZE succeeds in producing a realization if this is possible. However, we allow to alter or even eliminate some scheduling decisions if that should be necessary. When in Line 4 of Algorithm RE-ALIZE a route cannot be extended for ship i due to insufficient space for waiting in the preferred siding, we backtrack: Try the route extension from one siding earlier, dismissing the already fixed onwards part of the extension. This is iterated until we find a dynamic route for ship i or no earlier siding is available. In the latter case, *no* dynamic route can be assigned to ship i. Note that this backtracking does not touch other ships: a ship which originally succeeded ship i along a re-considered transit segment may change sequence with ship i. We therefore cannot guarantee that the route extension of Line 4 respects a given combinatorial frame in general. However, barring the mentioned complications it does. The backtracking version of Algorithm REALIZE is called Algorithm REALIZEb.

Combinatorial neighborhood. We would now like to improve existing solutions by local search, as summarized in Algorithm LS. Instead of working with our geometric encoding of an incumbent solution $\mathcal{P}$ directly, we define a neighborhood on the induced combinatorics $\zeta(\mathcal{P})$, slightly abusing notation. This fits well with our conclusions from Lemma 2.3.1 since it enables us to produce a much larger variety of schedules in the first place. A neighbor of $\zeta(\mathcal{P})$ is constructed by altering scheduling decisions on a single segment. Improvement heuristics on scheduling decisions or conflicts have also been used for train timetabling on a single track, see e.g., [CL95; Hig+97]. Figure 2.5 motivates how neighbors are constructed: every ship i waiting in a siding could switch precedence with (some of) the k conflicting ships it is waiting for. Let $j_1, \ldots, j_k$ be the ordering of these ships on E. All of them currently precede i on E. Every re-ordering $\{i, j_1, \ldots, j_k\}$, $\{j_1, i, j_2, \ldots, j_k\}$, $\ldots$, $\{j_1, \ldots, j_{k-1}, i, j_k\}$ induces new precedences $\rho'(E)$ on E. The union of all these, over all waiting ships in all sidings, describes the neighborhood of $\zeta(\mathcal{P})$. When several ships wait in a siding for the *same* set of conflicting ships, it makes sense to insert *all* of them simultaneously in $j_1, \ldots, j_k$. This is done in our implementation.

The described neighborhood may contain combinatorial frames that are not realizable, not even in the combinatorial relaxation of Section 1.3.2 that ignores siding capacities.

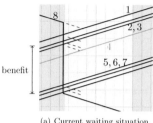
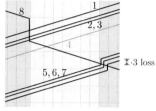

(a) Current waiting situation. (b) New realized schedule.

Figure 2.5: A waiting ship induces several neighbors in (a); a new incumbent solution in (b). This also illustrates how our local estimate of a neighbor is computed. Assume that ship 4 is not in conflict with ship 8. Each dashed line in (a) represents an opportunity for ship 8 to continue its journey. Realizing one of these would require swapping the scheduling relation for all succeeding ships on this transit segment. The most promising opportunity as in (b) yields a large benefit for ship 8 and three small increased waiting times for the ships 5, 6 and 7.

Theorem 2.1.1 ensures that we detect this defect, and an adapted Algorithm REALIZE can repair it. In Line 7 of Algorithm LS we ensure that there are free ships as long as there are ships with an incomplete route by heuristically deleting a relation on a cyclic dependency (similar to Figure 2.4a) when this is detected.

Local estimates of neighbors. Representing a solution only by its combinatorics adequately reflects our wish to improve scheduling decisions. At the same time, however, this also entails complications like the problem of how to evaluate a neighbor ζ': It would be too expensive to use Algorithm REALIZEb only to calculate the waiting time of a solution produced from ζ'. As a remedy we only use a simple estimate, see again Figure 2.5: We compare the decreased waiting time (benefit) of a ship, or a group of ships, with the increased waiting time (loss) of the conflicting ships we delay. Such local comparisons of ordering alternatives according to current arrival times and bundled decisions for close aligned trains are a natural idea. Cai and Goh [CG94] use similar considirations for a simple heuristic to fix orders within one iteration over time. Nevertheless, it is a very myopic analysis ignoring that reversing scheduling decisions locally can have global effects: because of siding capacities, waiting time can be saved or produced in a siding distant from the altered scheduling. Only neighbors with best local estimates are actually evaluated by trying to realize ζ' via Algorithm REALIZEb. Recall that we cannot guarantee that all altered scheduling decisions are respected, and that it may even happen that a ship cannot be routed at all. A solution with best actually realized improvement becomes the next incumbent. In fact, in order to escape local optima we allow worsening iterations. A parameter limits the total number of such failures of improvement. We avoid cycling by tabooing bad incumbents.

Heuristically encouraging certain properties of a solution. For reasons of practical acceptance of our solutions we need to take care of a few desired properties which have

---

input: canal C, requests R for ships S
output: solution $\mathcal{P}$
1 construct initial solution $\mathcal{P}$ by sequential routing
2 **repeat**
3     $\zeta(\mathcal{P}) :=$ combinatorics induced by $\mathcal{P}$
4   estimate benefits/losses of all neighbors of $\zeta(\mathcal{P})$
5     $\mathcal{N} :=$ subset of neighbors with best local estimates
6   **foreach** neighbor $\zeta' \in \mathcal{N}$:
7   ensure/repair (combinatorial) realizability of ζ' using Theorem 2.1.1
8   construct a solution from ζ' using Algorithm REALIZEb
9     $\mathcal{P} :=$ best found solution for candidates in $\mathcal{N}$
10 **until** there is not enough improvement
11 **return** globally best found solution

Algorithm LS: The general scheme of our local search on the combinatorics

not been algorithmically addressed so far. We therefore do not evaluate solutions only by their total waiting times for several reasons: (a) Algorithm REALIZEb may leave ships unrouted which nominally decreases waiting time; (b) very large individual waiting times of single ships may decrease that of many other ships, however in practice, waiting in a siding (and also the total waiting time per ship) should not exceed certain soft limits; (c) a large individual waiting time of a single ship counts as much as the sum of short waiting times for several ships; (d) it is a good idea to give priority to slow ships as they generally produce more scheduling complications in particular for opposed ships. These problems can be lessened by assigning individual weights to ships and by penalizing large individual waiting times. In our implementation this is done by replacing the objective of total waiting time by $\sum_{i \in S}(w_i W_i)^f$ with priorities w_i and $f > 1$. This applies to both, estimating a neighbor in Line 4 and choosing a best new incumbent in Line 9 of Algorithm LS.

2.3.3 Rolling horizon

Finally, we embed the local search in a rolling horizon framework, i.e., we consecutively plan for shorter, overlapping time horizons. There are two good reasons for doing so—an algorithmic and a conceptual one. Algorithmically, this supports the local search as more neighbors can be evaluated when dealing with a smaller instance i.e., a shorter time horizon. Conceptually, we more realistically emulate the planning process at the Kiel Canal where requests arrive over time, i.e., in an online manner.

We denote the *horizon length* by T_Δ, the *horizon start time* by T_0, and the *step size* by $T_\delta \le T_\Delta$. In each iteration we consider only a partial instance of those ships with release time between T_0 and the *horizon end time* $T_0 + T_\Delta$. Note that it is important to construct each dynamic route until the actual target position of the ship even if it will arrive there after $T_0 + T_\Delta$. After constructing a solution for this partial instance, we declare everything before the horizon end time as fixed and define the intersection point of each calculated dynamic route with the corresponding time axis as new start position and release time. Now, T_0 increments to $T_0 + T_\delta$ and the process repeats until all released ships have been considered.

Fixing everything before the horizon start time bears the risk that for some ships no dynamic route can be found since it can not be adapted if siding capacities are exceeded afterward. Besides the construction of dynamic routes to the actual target position beyond the scope of the horizon length we also transmit the already taken scheduling decisions ζ from one time horizon to the next to decrease this risk.

Experiments in Section 2.4.1 will show that this results in better objective values, running times and even in fewer ships without assigned dynamic route compared to plain local search.

2.4 Computational study

GPS position data are collected permanently for every vessel in the canal. The entire data (of about 43,000 ships) of actually traveled itineraries for the busy year 2007 were made available to us. From these we generated 365 instances, one for each day with a planning horizon of 24 hours. They contain 185 requests on average (minimum 83, maximum 247). Algorithms were implemented in Java. All computations for the following study were performed on a commodity desktop PC running Linux.

When evaluating the quality of a solution we refer to average waiting time per ship. The (penalty) waiting time of a ship that could not be routed by an algorithm is set to 2 hours. All calculations are exact to the second (typically not practicable in time-discretized approaches). Figure 2.6 shows the two visualization options we offer in our tool to validate our solutions. Both greatly helped in fine tuning and granting approval by the expert planners.

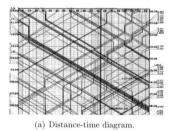
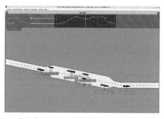

(a) Distance-time diagram. (b) Animation of ship movements.

Figure 2.6: Screen shots from our tool displaying solutions as (a) an interactive distance-time diagram or (b) an animation of the movement of ships on a map of the canal.

2.4.1 Algorithmic components

We stressed the interplay of geometry and combinatorics several times. Our first suite of experiments aims at demonstrating that this problem understanding also pays off computationally. We evaluate the "purely geometric" approach of sequential routing (Section 2.3.1, denoted SeqR); sequential routing integrated with re-scheduling decisions in a local search on the combinatorics (Section 2.3.2, denoted LS-240); and the rolling horizon heuristic which additionally reflects the problem's online character (Section 2.3.3, denoted RH-1). The numbers refer to the following parameter settings. LS-240 is allowed a maximal number of 240 worsening steps within the local search over the full 24-hour planning horizon. In RH-1 the horizon length is $T_\Delta := 2$ hours and the step size is $T_\delta := 1$ hour. As the local search acts only on horizons of an hour, we allow a maximum number of $10(= 240/24)$ worsening steps within the local search to make LS-240 comparable to RH-1 in that respect. Furthermore, we greedily evaluate only one promising neighbor-candidate within an improvement step, i.e., $|\mathcal{N}| = 1$ in Algorithm LS. The variant with $|\mathcal{N}| = 3$ is denoted by RH-3.

We present four pairwise comparisons of algorithms. In each comparison we compute, for each instance, the ratio of average waiting times obtained by the respective two algorithms and give the whole distribution of ratios as a box-and-whiskers plot in Figure 2.7. The box marks the range of the mid 50% of the ratios where the central line indicates the median. The whiskers mark the range of 95% and the smallest and largest 2.5% are considered as outliers, marked by circles. The average (marked with an asterisk $*$) and the standard deviation are given as information.

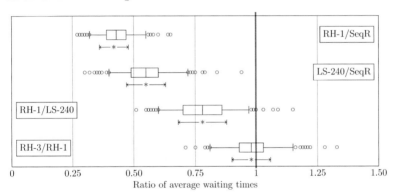

Figure 2.7: Four pairwise comparisons of algorithmic approaches: Each box-and-whisker plot shows the distribution of instance-wise ratios of average waiting times achieved by the respective two algorithms. A ratio $X/Y < 1$ means that X performs better than Y.

Both, the integration of routing and scheduling (LS-240) and the rolling horizon heuristic (RH-1) significantly improve over sequential routing (SeqR) alone. It is remarkable that the local search improvement performs considerably better when it works in small chunks "over time" instead of more globally on the whole day, as shown by the comparison RH-1/LS-240. The last comparison RH-3/RH-1 indicates that it does not pay much to evaluate neighbor-candidates in a way more elaborate than greedily. The most important qualitative conclusion is that taking care of scheduling decisions in addition to sequential routing is imperative for obtaining satisfactory solutions.

Table 2.1 summarizes two further aspects: the number of ships for which no dynamic route could be found (a little more than one ship per instance) and computation times (which are generally very low). RH-1 yields the best combination of quality and running time.

algorithm	# ships without route				running time (CPU min)			
	avg.	std. dev.	min	max	avg.	std. dev.	min	max
SeqR	1.26	1.36	0	9	0.01	0.01	0.00	0.05
LS-240	1.10	1.30	0	8	7.97	3.76	0.21	27.80
RH-1	1.05	1.22	0	6	1.34	1.42	0.23	19.65
RH-3	1.05	1.22	0	7	3.56	2.77	0.15	34.40

Table 2.1: Performance of algorithmic variants, averaged over all instances: Number of ships for which no dynamic route could be found and the running time in CPU minutes.

2.4.2 Combinatorial relaxation

In Section 1.3.2 we discussed a combinatorial relaxation that ignores siding constraints. We will use its MIP formulation (1.4)–(1.11) to estimate the quality of solutions obtained by our rolling horizon heuristic RH-1. This MIP is computationally very demanding: a still unsolved instance (called `shipsched`) with 87 ships belongs to the *challenge* benchmark set of MIPLIB2010 [Koc+11]. We therefore prepared smaller instances comparable to one time horizon in RH-1 as follows. Start from a given solution for a whole day; filter all ships that are present or enter the canal during a given time horizon of two hours; and further reduce the number of ships to at most 35 by random removal, biased to prefer a balance between rightbound and leftbound ships and to keep ships satisfying a minimum distance value between start and target position.

Figure 2.8 shows results for 21 instances. Label "MIP" marks the value of the best known feasible integer solution of MIP (1.4)–(1.11). For instances not solved to optimality we additionally state the largest known lower bound indicated with the label "LB." The objective values labeled by "RH-1" refer to the rolling horizon heuristic. To add another level of comparability we also consider a version of the rolling horizon heuristic where siding constraints are relaxed as in the MIP; the corresponding label is "relaxRH-1". As an information we state the computation times needed by the state-of-the-art MIP solver `CPLEX` (version 12.3). They indicate that the complexity of instances of 35 ships varies from very easily solvable to intractable.

The average difference of waiting times between LB/MIP and relaxRH-1 is slightly above 5 minutes (maximum is 18 minutes). These gaps increase a bit if the siding conditions must be respected (average below 7 minutes, maximum below 22; ignoring the two instances marked with ∗). One might have expected a larger increase. There are even two instances where RH-1 produces less waiting time than relaxRH-1. In these cases, siding capacity was not critical but induced different decisions within the local search process.

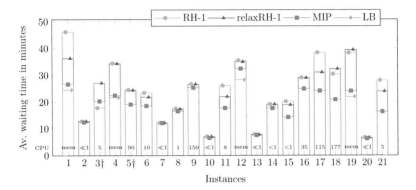

Figure 2.8: Results for 21 small instances with at most 35 ships over a 2h horizon. The columns indicate the respective average waiting times per instance for the rolling horizon heuristic (RH-1), this heuristic with relaxed siding conditions (relaxRH-1), the value of the best known integer solution of MIP (1.4)–(1.11), and (if not identical to MIP) the largest known lower bound (LB). MIP/LB constitute *lower* bounds on the optimum. A dagger † indicates that RH-1 was not able to find a dynamic route for all ships of the corresponding instance due to capacity problems within sidings. The CPLEX computation time to solve the MIP is stated in CPU minutes, or "mem" when the computation hit the memory limit.

Returning to the 24-hour instances, this effect is still visible but remains exceptional, see Figure 2.9. On average, RH-1 loses only 8% when compared to relaxRH-1, but this increases drastically (to about 30%) for the prognosticated traffic demand. This shows the importance of modeling the siding constraints in detail.

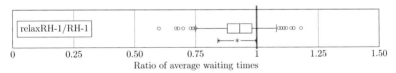

Figure 2.9: Comparison as in Figure 2.7 (all instances of 2007): our rolling horizon heuristic (RH-1) vs. its variant with relaxed siding constraints (relaxRH-1).

2.4.3 GPS data realized

We finally evaluate the practical usefulness of our algorithms. The GPS data contain actual entry/exit times at sidings for each ship. Figure 2.10(a) shows the linear interpolation between them. Small circles indicate infeasibilities of such a solution: large ships pass each other (slightly) outside sidings; velocities do not exactly match the actual ones (some ships exceed the speed limit in reality, others move below full speed, in particular when waiting in front of the lock chambers at the canal boundaries); also constraints in sidings may be violated.

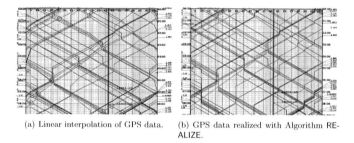

(a) Linear interpolation of GPS data. (b) GPS data realized with Algorithm RE-ALIZE.

Figure 2.10: Actually traveled itineraries as per GPS data with infeasibilities on the left (a) and a feasible solution derived using Algorithm REALIZE on the right (b).

We deduce a combinatorial frame from this solution, resorting to the most plausible scheduling decision in case of ambiguities, and use Algorithm REALIZE to construct a routing. The latter may not *exactly* realize the original solutions since these need not be earliest arrival ones in general. However, we consider this proceeding a closest-to-reality re-construction of actual itineraries which at the same time fit the precise feasibility definitions. The result is called GPS data realized, GPS-re.

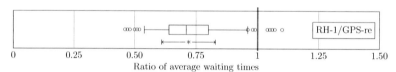

Figure 2.11: Distribution of the instance-wise improvement w.r.t. average waiting time of the rolling horizon heuristic (RH-1) over realized GPS data (GPS-re).

Figure 2.11 shows the improvement of the rolling horizon heuristic over manually planned solutions. Again, precise values should be taken with care, but a tenor is clearly visible. More interesting is the behavior pointed out in Figure 2.12. Instances are sorted according to average waiting time which is considered a proxy for the *planning complexity*. Instances which are harder under this measure in reality are harder for our heuristic, too. This similarity brings us in a position to reliably emulate predicted traffic under different canal enlargement options.

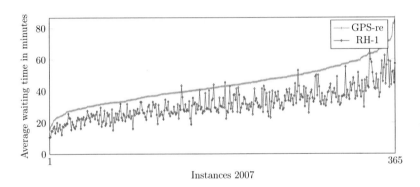

Figure 2.12: Average waiting time achieved by realizing the GPS data (GPS-re) and the rolling horizon heuristic (RH-1) for each day of 2007, sorted non-decreasingly according to GPS-re.

3

Offline Complexity
of Bidirectional Scheduling

This chapter investigates the computational complexity of bidirectional scheduling in the offline setting. A straightforward reduction from single machine scheduling yields immediately the NP-hardness for bidirectional scheduling which is induced by the possible variation of processing times. Nevertheless, from the application point of view the transit times are dominating the processing times and hence we are especially interested how to deal with delays induced by opposed jobs. Therefore, we investigate the simplification where all jobs are identical. NP-hardness results are complemented by restrictions admitting efficient exact algorithms. Therefore, we understand which properties apart from processing times cause difficulties.

This chapter is based on joint work with Yann Disser and Max Klimm [Dis+15].

The bidirectional scheduling model proposed in Section 1.4 captures the essence of bidirectional traffic by distinguishing processing and transit times. Within this chapter, we use this simple framework to exhibit the computational key challenges of bidirectional traffic. For the special case of one segment, jobs only traveling from left to right, and zero transit times, the problem corresponds to total completion time minimization in non-preemptive single-machine scheduling with release dates. From this special case we directly conclude NP-hardness of bidirectional scheduling, cf. Lenstra et al. [Len+77].

Theorem 3.0.1 (Lenstra et al. [Len+77]). *Minimizing the total completion time for bidirectional scheduling is* NP-*hard even if $m = 1$.*

For our applications in mind it often takes much longer for a vehicle to traverse a segment then to enter it. From this point of view we expect the transit times to dominate the processing times. That's why we are especially interested how to deal with delays induced by opposed jobs. Hence, we investigate the simplification where all jobs are identical concerning processing times and weights. The case of one segment where no two jobs are compatible can be solved by dynamic programming since the jobs of each heading can be scheduled in order of their release leaving the remaining open decision of where to switch between the two directions. Ensuing from this, we can increase the complexity in two directions: we can consider more complex compatibilities or multiple segments. Table 3.1 gives an overview of our results.

	—————— Number m of segments ——————		
compatibilities	$m = 1$	m const.	m arbitrary
Identical jobs $p_{ij} = p$, $\tau_{ij} = \tau_i$			
none	polynomial	polynomial[3]	NP-hard[1]
const. # of types	[Thm. 3.3.1]	[Thm. 3.3.2]	[Thm. 3.1.1]
arbitrary	NP-hard[2] [Thm. 3.2.5]		NP-hard[1,2]
Different jobs $p_{ij} = p_j$, $\tau_{ij} = \tau_i$			
uniform	NP-hard [Len+77]		NP-hard[1]

Table 3.1: Overview of total completion time complexity for bidirectional scheduling.

We first show that the BSP is NP-hard if the number of segments is not restricted, even without processing times and with identical transit times (Section 3.1). Therefore, we get a difference to one-directional (flow shop) scheduling with identical processing times, which is trivial. The hardness proof is via a non-standard reduction from MAX-CUT. The key challenge is to use the local interaction of the jobs on the path to model global interaction between the vertices in the MAX-CUT. We overcome this issue by introducing polynomially many vertex gadgets encoding the partition of each vertex and synchronizing these copies along the instance.

Back on a single segment, we show NP-hardness for arbitrary compatibilities, even with unit processing and transit times (Section 3.2). We use a reduction from a variant of SAT with a sparse compatibility graph that only allows for specific combinations of jobs to run concurrently.

Reducing both complexities at the same time finally allows for exact algorithms (Section 3.3). On a single segment there is a polynomial time dynamic program if we assume that the jobs can be clustered into a constant number of compatibility types. For more than one segment this kind of dynamic program is not applicable. But we present an enumeration algorithm that has polynomial running time for sufficient assumptions on the given numbers. This works in particular with zero processing times and unit transit times. Hence, this polynomial solvable case with a bounded number of segments complements the NP-hard case with a path of unbounded length.

The achieved insights are also of practical relevance. Bidirectional infrastructures are often fixed buildings by nature. The Kiel Canal has no more than 12 sidings. Also railway networks do not vary heavily over time. Still, the occurring numbers should be tractably small to make use of this. The compatibilities constitute in this simple model the main difference between train time tabling and ship traffic control. In that sense, also a difference in complexity can be observed. Nevertheless, a small number of different vessels helps to increase tractability. If for example the compatibilities are modeled by the simple criteria of fixed traffic groups and passage numbers we get a constant number of compatibility types. If applying all additional constraints necessary for the Kiel Canal it gets again more involved.

---

[1] even if $p = 0$, $\tau_i = 1$
[2] even if $\tau_i = p = 1$
[3] only if $p = 1$, $\tau_i \leq$ const, $r_j, \tau_i \in \mathbb{N}$

3.1 Hardness for multiple segments

We start to consider a path of arbitrary length and show that the BSP is hard, even when all processing times are zero and all transit times coincide. In other words, we eliminate all interaction between jobs in the same direction and show that hardness is merely due to the decision when to switch between left- and rightbound operation of each segment. Formally, we show the following result.

Theorem 3.1.1. *Minimizing the total waiting time for bidirectional scheduling is* NP-*hard even if* $p_j = 0$ *and* $\tau_i = 1$ *for each* $j \in J$ *and* $i \in M$.

We reduce from the MAX-CUT problem which is contained in Karp's list of 21 NP-complete problems [Kar72].

MAX-CUT

Given: An undirected graph $G = (V, E)$ and $k \in \mathbb{N}$.
Question: Is there a partition $V = V_1 \dot\cup V_2$ with $|E \cap (V_1 \times V_2)| \geq k$?

Let an instance $\mathcal{I} = (G_{\mathcal{I}}, k)$ of MAX-CUT be given, with $G_{\mathcal{I}} = (V_{\mathcal{I}}, E_{\mathcal{I}})$, $|V_{\mathcal{I}}| = n_{\mathcal{I}}$, and $|E_{\mathcal{I}}| = m_{\mathcal{I}}$. We introduce a set of jobs on polynomially many segments that can be scheduled with a total waiting time of W if and only if $\mathcal{I}$ admits a solution. Our construction is comprised of various gadgets which we describe in the following. We make use of suitably large parameters $x \gg y \gg z \gg 1$ that we will specify later. For example, x is chosen in such a way that if ever x jobs are located at the same segment, these jobs need to be processed immediately in order to achieve a waiting time of W. Note that because jobs take no time in being processed (i.e., $p_j = 0$), we can schedule any number of jobs sharing direction simultaneously on a single segment. Also, since $\tau = 1$, it makes no sense for a segment to stay idle if jobs are available. This allows us to restrict our analysis to schedules that are *sensible* in the sense that for each segment and at every time step all jobs in one direction available at the segment get scheduled. On the other hand, the non-zero transit time induces a cost of switching the direction of jobs that are processed at a segment.

Vertex Representation. A cornerstone of our construction is the *vertex gadget* that occupies a fixed time interval on a single segment and can only be (sensibly) scheduled in two ways (cf. Figure 3.5), which we interpret as the choice whether to put the corresponding vertex in the first or second part of the partition, respectively. We introduce multiple *vertex segments* that each have exactly one vertex gadget for each vertex in $\mathcal{I}$ and add further gadgets that ensure that the state of all vertex gadgets for the same vertex is the same across all segments.

Vertex Gadget. Each of the segments $1, 10, 19, 28, \ldots$ hosts one *vertex gadget* for each of the vertices in $V_{\mathcal{I}}$ (cf. Figure 3.1 with the following). Each vertex gadget g_t on segment $9\ell + 1$ occupies a distinct time interval $[13t, 13(t+1))$, $t < n_{\mathcal{I}}$, on the segment and is associated with one of the vertices $v \in V_{\mathcal{I}}$. The gadget comes with $24y$ *vertex jobs* that only need to be processed at segment $9\ell + 1$, half of them being leftbound, half being rightbound. Exactly y jobs of each direction are released at times $13t, 13t + 1, \ldots, 13t + 11$. We say that g_t is scheduled *consistently* if either all leftbound vertex jobs are processed immediately when

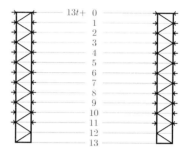

Figure 3.1: Illustration of the vertex gadget in the leftbound (left) and the rightbound (right) state. At each time $13t, 13t + 1, \ldots, 13t + 11$ multiple right- and leftbound jobs are released. Since all jobs have processing time 0, jobs in the same direction can be processed simultaneously. The only two sensible schedules differ in whether leftbound jobs are processed at even or odd times.

they are released and all rightbound jobs wait for one time unit, or vice-versa. We say the gadget is in the *leftbound* (*rightbound*) *state* and interpret this as vertex v being part of set V_1 (V_2) of the partition of $V_{\mathcal{I}} = V_1 \cup V_2$ we are implicitly constructing. A schedule is *consistent* if all vertex gadgets are scheduled consistently.

The following lemma allows us to distinguish consistent schedules.

Lemma 3.1.2. *The vertex jobs of a single vertex gadget can be scheduled consistently with a waiting time of $12y$, while every inconsistent schedule has waiting time at least $13y$.*

Proof. Since $p = 0$, we can schedule all available jobs with the same direction simultaneously. It follows that both consistent schedules are valid, and, since in both exactly half of the vertex jobs wait for one unit of time, the total waiting time of such a schedule is $12y$. Any inconsistent (sensible) schedule would have to send jobs in the same direction in two consecutive unit time intervals, which means that in addition to the minimum waiting time of $12y$, at least y jobs have to wait an extra unit of time. □

Synchronizing Vertex Gadgets. Since every vertex $v \in V_{\mathcal{I}}$ is represented by multiple vertex gadgets on different segments, we need a way to ensure that all vertex gadgets for v are in agreement regarding which part of the partition v is assigned to. We introduce two different gadgets that handle synchronization. These gadgets allow us to synchronize vertex gadgets on consecutive vertex segments in two ways. We can either simply synchronize vertex gadgets that occupy the same time interval on the two vertex segments (*copy gadget*), or we can synchronize pairs of vertex gadgets occupying the same consecutive time intervals on the two vertex segments by linking the first gadget on the first segment with the second one on the second segment and vice-versa, i.e., we can transpose the order of two consecutive gadgets from one vertex segment to the next (*transposition gadget*). Using a combination of copy and transposition gadgets, we can transition between any two orders of vertex gadgets on distant segments.

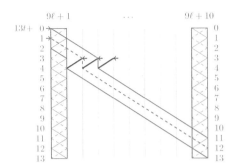

Figure 3.2: Illustration of the copy gadget between two vertex gadgets. All sensible trajectories of the synchronization jobs are displayed, assuming that the vertex gadgets are in the same state (rightbound state solid vs. leftbound state dashed).

Copy Gadget. The *copy gadget* synchronizes the vertex gadgets g_t on two segments $9\ell + 1$ and $9\ell + 10$ (cf. Figure 3.2 with the following). The gadget consists of $2z$ rightbound *synchronization jobs*, half of which are released at time $13t$ and half at time $13t + 1$. Each synchronization job travels from start segment $9\ell + 1$ to target segment $9\ell + 10$. In addition, we introduce $3x$ *blocking jobs* that are used to enforce that specific time intervals on a segment are reserved for leftbound/rightbound operation. Essentially, releasing x blocking jobs at time t on a single segment prevents any jobs to run in opposite direction during the time interval $[t, t + 1)$ (and even earlier). In this manner, we block the interval starting at time $13t + 3$ on segments $9\ell + 2, 9\ell + 3, 9\ell + 4$.

Lemma 3.1.3. *In any consistent schedule, the synchronization jobs of a single copy gadget can be scheduled with a waiting time of $3z$ if the two corresponding vertex gadgets are in the same state, otherwise their waiting time is at least $5z$.*

Proof. Since $x \gg z$, we need to schedule all blocking jobs as soon as they are released. If both vertex gadgets g_t linked by the copy gadget are in the rightbound state, the synchronization jobs released at time $13t$ only have to wait for one time unit at segment $9\ell + 4$, while the other jobs have to wait at segments $9\ell + 1$ and $9\ell + 2$. Similarly, if the vertex gadgets are in the leftbound state, the first half of the jobs have to wait at segments $9\ell + 1$ and $9\ell + 3$, while the other half only has to wait at segment $9\ell + 3$. The waiting time in either case is $3z$. If the vertex gadgets are in opposite states, all jobs have to additionally wait until segment $9\ell + 10$, which results in a total waiting time of at least $5z$. $\qquad\square$

Transposition Gadget. The transposition gadget synchronizes the vertex gadgets g_t, g_{t+1} on segment $9\ell + 1$ with the vertex gadgets g_{t+1}, g_t on segment $9\ell + 10$ (cf. Figure 3.3 with the following). The challenge here is that jobs synchronizing the different pairs of vertex gadgets need to pass each other without interfering. We achieve this by making sure that the jobs never meet while being in transit at the same segment. The gadget consists of $4z$ synchronization jobs, half being rightbound and half being leftbound. Half of each are released at times $13t + 6$ and $13t + 7$, and all need to be processed at segments $9\ell +$

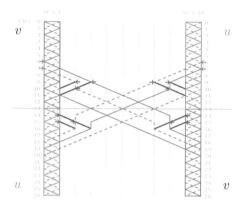

Figure 3.3: Illustration of the transposition gadget. The dashed lines depict all sensible trajectories of the synchronization jobs, assuming that the vertex gadgets are pairwise in the same states. Note that jobs in different directions never meet while in transit through the same segment.

$1, \ldots, 9\ell + 10$ (in different directions). In addition, we introduce $12x$ blocking jobs to block the intervals starting at the following times: at times $13t + 9$, $13t + 10$ for rightbound jobs and at times $13t + 14$, $13t + 15$ for leftbound jobs on segment $9\ell + 2$, at times $13t + 9$ for rightbound and at $13t + 15$ for leftbound on segment $9\ell + 3$, and the corresponding (symmetrical) intervals in opposite direction on segments $9\ell+8$ and $9\ell+9$ (cf. Figure 3.3).

Lemma 3.1.4. *In any consistent schedule, the synchronization jobs of a single transposition gadget can be scheduled with a waiting time of $10z$ if each of the two pairs of corresponding vertex gadgets are in the same state, otherwise their waiting time is at least $12z$.*

Proof. Since $x \gg z$, we need to schedule all blocking jobs as soon as they are released. It is easy to verify that all synchronization jobs wait for exactly 2 time units due to blocking jobs. In addition, half of the jobs wait for one unit of time at the segment where they are released – for a total of $10z$ time units. If the pair of vertex gadgets is in opposite states, all connecting synchronization jobs need to wait at least one additional unit of time at their last segment. Observe that synchronization jobs in opposite directions are never in transit on the same segment at the same time. □

Edge Representation. We construct an edge gadget for each edge of $\mathcal{I}$ that incurs a small waiting time if the two corresponding vertex gadgets in consecutive time intervals and segments are in different states and a slightly higher waiting time if they are in the same state. By tuning the multiplicity of each job, we can ensure that only schedules make sense where vertex gadgets are scheduled consistently. Minimizing the waiting time then corresponds to maximizing the number of edge gadgets that link vertex gadgets in different states, i.e., maximizing the size of a cut.

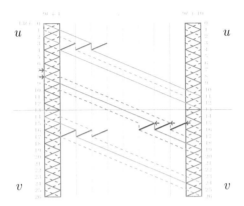

Figure 3.4: Illustration of the edge gadget. All sensible trajectories of the edge jobs are displayed, assuming that the vertex gadgets are in opposite states. Note that edge jobs do not interact with synchronizing jobs of copy gadgets for both vertices.

Edge Gadget. For any two vertices u, v incident in $G_{\mathcal{I}}$ we will implement an *edge gadget* between vertex gadget g_t on segment $9\ell + 1$ representing u and vertex gadget g_{t+1} on segment $9\ell + 10$ representing v. The edge gadget itself consists of 2 rightbound *edge jobs*, one being released at time $13t + 7$ and the other at time $13t + 8$. Both jobs need to be processed on segments $9\ell + 1, \ldots, 9\ell + 10$. We add $3x$ blocking jobs to block the unit time interval starting at time $13t + 15$ on segments $9\ell + 7, 9\ell + 8, 9\ell + 9$.

Lemma 3.1.5. *In any consistent schedule, the edge jobs of a single edge gadget can be scheduled with a waiting time of 3 if the two connected vertex gadgets are in opposite states, otherwise their waiting time is at least 5.*

Proof. One job always has to wait for a time unit at the first segment. Both jobs have to wait for the blocking jobs (since $x \gg 1$). If the vertex gadgets are in the same state, both jobs have to wait an additional unit of time at the last segment. □

Construction. In order to fully encode the given MAX-CUT instance $\mathcal{I}$, we need to introduce an edge gadget for each edge in $E_{\mathcal{I}}$. However, edge gadgets can only link vertex gadgets in consecutive time intervals. We can overcome this limitation by adding a sequence of vertex segments and transposing the order of two vertex gadgets from one segment to the next as described before. With a linear number of vertex segments we can reach an order where the two vertex gadgets we would like to connect with an edge gadget are adjacent. At that point, we can add the edge gadget, and then repeat the process for all other edges in $\mathcal{I}$ (cf. Figure 3.5).

Proof of Theorem 3.1.1. We start by introducing a vertex gadget g_t on segment 1 for each vertex $v_t \in V_{\mathcal{I}}$ of the given MAX-CUT instance $\mathcal{I}$. For each edge $\{u, v\} \in E_{\mathcal{I}}$ we extend the construction by appending more segments as follows. We add a sequence of blocks of 9 segments, the last of which contains again a vertex gadget for each vertex. In between we

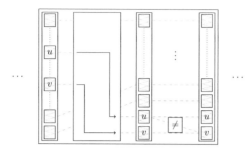

Figure 3.5: Illustration of our hardness construction for a single edge $e = \{u, v\}$. First, a sequence of segments is used to change the order of vertex gadgets, such that the vertex gadgets corresponding to u and v occupy consecutive time intervals. Then, an edge gadget is added that incurs an increased waiting time if the vertex gadgets for u and v are in the same state.

add copy and transposition gadgets in such a way that on the last segment i the vertex gadgets g_0 and g_1 represent the vertices u and v. We can achieve this by adding less than $n_{\mathcal{I}}$ blocks. We add an additional block of 9 segments, and add copy gadgets for each of the variables. Finally, we add an edge gadget connecting vertex gadget g_0 on segment i with g_1 on the last segment. Observe that the edge jobs do not interfere with any of the synchronization jobs for the copy gadgets for the first two vertices (cf. Figure 3.4). We repeat the process once for each edge. The total number of blocks is then bounded by $n_{\mathcal{I}}m_{\mathcal{I}}$. The number of vertex gadgets is $n_v < n_{\mathcal{I}}^2 m_{\mathcal{I}}$, and the number of transposition and copy gadgets is $n_t < n_c < n_v$. Hence, we have a total number of segments in $\mathcal{O}(n_{\mathcal{I}}m_{\mathcal{I}})$, and a total number of jobs in $\mathcal{O}(n_{\mathcal{I}}^2 m_{\mathcal{I}}(x + y + z))$.

We claim that if the MAX-CUT instance admits a solution $\mathcal{S}$, we can schedule all jobs with waiting time at most $W = 12n_v y + 3n_c z + 10n_t z + 5m_{\mathcal{I}} - 2k$. We do this by scheduling all vertex gadgets consistently in the state corresponding to the part of the partition the corresponding vertex belongs to in $\mathcal{S}$. Lemmas 3.1.2 through 3.1.4 guarantee that we can schedule everything but the edge jobs without incurring a waiting time greater than $12n_v y + 3n_c z + 10n_t z$. Finally, since at least k edges in the MAX-CUT solution are between vertices in different sets of the partition, and the vertex gadgets are set accordingly, by Lemma 3.1.5, we obtain an additional waiting time of at most $5m_{\mathcal{I}} - 2k$ as claimed.

It remains to establish that the waiting time exceeds W in case the MAX-CUT instance does not admit a solution. We set $x = W + 1$, such that all blocking jobs have to be scheduled as soon as they are released. By Lemma 3.1.2, scheduling at least one vertex gadget inconsistently produces a total waiting time of at least $12n_v y + y$. We now set $y = 18n_{\mathcal{I}}^2 m_{\mathcal{I}} z > 3n_c z + 10n_t z + 5m_{\mathcal{I}}$ for the vertex jobs, such that a single inconsistent vertex gadget results in a waiting time greater than W. Hence, each vertex gadget needs to be scheduled consistently. By Lemmas 3.1.3 and 3.1.4, we have that if not all vertex gadgets corresponding to the same vertex are in the same state, the waiting time for vertex and synchronization jobs is at least $12n_v y + 3n_c z + 10n_t z + z$. We set $z = 5m_{\mathcal{I}}$, which allows us to conclude that all vertex gadgets are in agreement regarding the partition of the vertices. Finally, Lemma 3.1.5 enforces that there are at least k edge gadgets between vertices in different states. This however is impossible as our MAX-CUT instance does not admit a solution. □

We can reformulate Theorem 3.1.1 for nonzero processing times, simply by making the transit time large enough that the processing time does not matter.

Corollary 3.1.6. *Minimizing the total waiting time for bidirectional scheduling is* NP-*hard even if $p_j = 1$ and $\tau_i = \tau$ for each $j \in J$ and $i \in M$.*

Proof. We adapt our construction by setting $p = 1$ and $\tau = n^2 m$ and scaling all release times by $n^2 m$, where n, m are the number of jobs and segments, respectively. We claim that the original instance admits a solution of some waiting time W if and only if it now admits a solution with waiting time in $[W\tau, (W + 1)\tau)$. This proves the Corollary, as the intervals are pairwise disjoint for different (integer) values of W.

If the original construction (with $p = 0$ and $\tau = 1$) does not admit a solution with waiting time at most W, then a scaled version with $p = 0$ and $\tau = n^2 m$ does not admit a solution with waiting time at most $W\tau$. But the lowest possible waiting is monotonically increasing with increasing processing times, hence the adapted instance with $p = 1$ does not admit a solution of waiting time at most $W\tau$.

Conversely, assume we have a solution of the original instance with waiting time W. We fix the order in which jobs are processed along each segment and construct a schedule for the setting $p = 1$, $\tau = n^2 m$ by introducing additional waiting periods for each job. Clearly, each job has to wait at most one time unit for each other job to be processed at each segment. Hence, the additional waiting time overall is smaller than $n^2 m = \tau$. □

3.2 Hardness of custom compatibilities

After establishing the complexity of an unbounded number of segments we now turn back to a single segment and investigate the influence of compatibilities. For arbitrary compatibility graphs we show that bidirectional scheduling is NP-hard already on a single segment with unit processing and transit times. For ease of exposition, we first consider the minimization of the makespan and then extend our result to minimum completion time.

3.2.1 Makespan minimization

Theorem 3.2.1. *Minimizing the makespan for bidirectional scheduling on a single segment with an arbitrary compatibility graph G_1 is* NP-*hard even if $p_j = \tau_1 = 1$ for each $j \in J$.*

We give a reduction from an NP-hard variant of SAT (cf. [GJ79]). Note the difference to the polynomially solvable $(3, 3)$-SAT, where each variable appears in *exactly* three clauses [Tov84].

(≤3,3)-SAT

Given: A formula with a set of clauses C of size three over a set of variables X, where each variable appears in at most three clauses.

Question: Is there a truth assignment of X satisfying C?

For a given $(\leq 3, 3)$-SAT formula we construct a bidirectional scheduling instance that can be scheduled within some specific makespan $C_{\max}$ if and only if the given formula is satisfiable. We partition the time horizon into 5 parts $P_1, \ldots, P_5$ with start time $A_1 = 0, A_2 = 6|X|, A_3 = 10|X|, A_4 = 10|X| + 2|C|$, and $A_5 = 12|X| + |C|$. The demanded makespan $C_{\max} = A_5 + 1$ will enforce that all jobs start before the end of the fourth part.

Confer Figure 3.6 along with the following overview. In all figures of this proof, time is directed downwards, and all rightbound jobs are depicted to the left and all leftbound jobs to the right of the segment. Since compatible jobs can run concurrently, the schedules of the leftbound and the rightbound jobs are drawn separately.

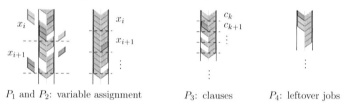

| P_1 and P_2: variable assignment | P_3: clauses | P_4: leftover jobs |

Figure 3.6: Illustration of the four parts of our construction put side by side.

The rough idea is as follows: In the first four parts we release a tight frame of *blocking jobs* B and *dummy jobs* H that need to be scheduled at their release date in any schedule that achieves $C_{\max}$. We can enforce this by making sure that at least one blocking job is released at (almost) each unit time step and that blocking jobs that are not supposed to run concurrently are incompatible. We use these jobs to create gaps for *variable jobs* that represent the variable assignments. By defining the compatibilities for the blocking jobs we are able to control which of these assignment jobs can be scheduled into each gap. In the first part of our construction, we release all variable jobs, which come in two *types*: one type representing a *true* assignment to the corresponding variable and the other type representing a *false* assignment. Our construction will enforce the following properties in each of its parts:

Lemma 3.2.2.   *In every feasible schedule with makespan $C_{\max}$, all jobs released before A_3 are scheduled in parts P_1 and P_2, except for two rightbound variable jobs of same type for each variable.*

Lemma 3.2.3.   *In every feasible schedule with makespan $C_{\max}$, the only jobs released before A_3 and scheduled in P_3 are rightbound variable jobs, where each corresponds to a variable assignment satisfying a different clause.*

Lemma 3.2.4.   *In every feasible schedule with makespan $C_{\max}$, the only jobs released before A_4 and scheduled in P_4 are rightbound variable jobs, and there are not more than $2|X| - |C|$ of them.*

In the following we explicitly define the released jobs of each part achieving the above properties. Each part is accompanied by a more detailed figure illustrating when jobs are released, the respective compatibility graph and an example of a schedule.

We start by specifying the jobs released in P_4.

Jobs of P_4. In the last part, $2|X|-|C|$ leftbound blocking jobs $B_4 = \{b_i \mid i = 0, \ldots, 2|X| - |C| - 1\}$ are released at $A_4 + i$ for each $b_j \in B_4$ in part P_4 leaving space for leftover rightbound variable jobs not scheduled until the beginning of this part. Each of the blocking jobs is only compatible with all rightbound variable jobs.

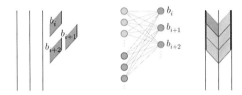

Figure 3.7: Part P_4 with blocking jobs reserving space for all remaining rightbound variable jobs.

Proof of Lemma 3.2.4. First, observe that with the required makespan of $A_5 + 1 = A_4 + 2|X| - |C| + 1$ each blocking job of B_4 must be scheduled directly at its release date. Consequently, there is no room to delay the start of any leftbound job released before P_4 to this part. Due to the compatibilities, the rightbound blocking and dummy jobs released before P_4 are also forced to run before the start of P_4. Therefore, there are exactly $2|X| - |C|$ open slots within P_4 reserved for rightbound variable jobs. $\square$

Jobs of P_3. The third part (Figure 3.8) is responsible for the assignment of satisfying literals to each clause. It consists of a set of jobs B_3 containing one leftbound blocking job b_k per clause c_k released at $A_3 + 2k$, which is compatible with each rightbound variable job that represents a variable assignment satisfying this clause. The gaps between the blocking jobs are filled with dummy jobs H_3 containing one rightbound job h_k^r and one leftbound job h_k^l with release date $A_3 + 2k + 1$ per clause $c_k \in C$. Each leftbound dummy job is compatible with all rightbound variable jobs, furthermore each rightbound dummy job h is compatible with the three leftbound jobs released in $[r_h - 1, r_h + 1]$.

Proof of Lemma 3.2.3. By Lemma 3.2.4 all jobs released within P_3 must start before the end of P_3. Hence, each leftbound dummy and blocking job is forced to start at its release date. Therefore, due to the compatibilities, each rightbound dummy job must be scheduled directly when released. The only remaining $|C|$ free slots can be filled with rightbound variable jobs – exactly one free slot per clause c_k reserved for a variable job representing an assignment that satisfies c_k. $\square$

We continue by introducing the released jobs within P_1 and P_2 to prove Lemma 3.2.2 afterward.

Jobs of P_1. In the first part, we release different kinds of jobs per variable (cf. Figure 3.9). There are rightbound variable jobs $T^r = \{t_{i,1}^r, t_{i,2}^r \mid x_i \in X\}$ representing a true assignment as well as rightbound variable jobs $F^r = \{f_{i,1}^r, f_{i,2}^r \mid x_i \in X\}$ representing a false

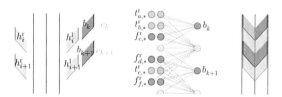

Figure 3.8: Part P_3 for $c_k = (x_a \vee x_b \vee \bar{x}_c)$ and $c_{k+1} = (\bar{x}_d \vee x_e \vee \bar{x}_f)$. Note that each variable job can be adjacent with more than one clause job (although this does not occur in the example).

assignment. These jobs are complemented by leftbound variable jobs $F^l = \{f_i^l \mid x_i \in X\}$ and $T^l = \{t_i^l \mid x_i \in X\}$ and further leftbound jobs $Q = \{q_i^t, q_i^f \mid x_i \in X\}$ called *indefinite*, both with the purpose to enforce a consistent assignment. To do so, we implement a certain structure by leftbound jobs $B_1 = \{b_i^t, b_i^f \mid x_i \in X\}$ for blocking, and some further dummy jobs $H_1 = \{h_i^{rt}, h_i^{lt}, h_i^{rf}, h_i^{lf} \mid x_i \in X\}$ for filling, for each variable and each value one leftbound and one rightbound true job. We release the rightbound true jobs $t_{i,1}^r$ at $6i$ and $t_{i,2}^r$ at $6i + 1$, the rightbound false jobs $f_{i,1}^r$ at $6i + 3$ and $f_{i,2}^r$ at $6i + 4$. Each indefinite job q_i^t together with the leftbound f_i^l is released at $6i + 1$, each q_i^f together with t_i^l at $= 6i + 4$. Furthermore we release the blocking jobs b_i^t at $6i$ and b_i^f at $6i + 3$ as well as the dummy jobs h_i^{rt}, h_i^{lt} at $6i + 2$ and h_i^{rf}, h_i^{lf} at $6i + 5$. The compatibility graph G_1 is defined such that each blocking job b_i^t is compatible with the corresponding $t_{i,1}^r$ and $t_{i,2}^r$, and each b_i^f with $f_{i,1}^r$ and $f_{i,2}^r$, respectively. The first indefinite job q_i^t is compatible with the corresponding rightbound true jobs t_i^r and t_i^r as well as the second q_i^f with f_i^r and f_i^r, respectively. Finally, we define each dummy job $h \in H_1$ to be compatible with the opposed jobs released in $[r_h - 1, r_h + 1]$. None of the remaining pairs of jobs are compatible.

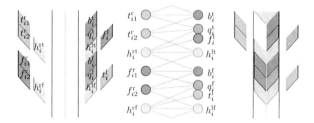

Figure 3.9: Released jobs per variable x_i in P_1, the corresponding compatibilities given by G_1 and a scheduled example for a true variable assignment.

Jobs of P_2. In the second part (Figure 3.10), there is room for exactly one indefinite job and one leftbound variable job per variable. This is realized by a set of rightbound blocking jobs $B_2 = \{b_{i,1}, b_{i,2} \mid x_i \in X\}$ where each $b_{i,1}$ released at $A_2 + 4i$ is compatible with the corresponding two indefinite jobs q_i^t and q_i^f. Each $b_{i,2}$ released at $A_2 + 4i + 2$ is compatible with the corresponding two leftbound variable jobs f_i^l and t_i^l. The gaps between two subsequent released blocking jobs are closed in both directions by dummy jobs $H_2 = \{h_{i,1}^{\mathrm{r}}, h_{i,2}^{\mathrm{r}}, h_{i,1}^{\mathrm{l}}, h_{i,2}^{\mathrm{l}} \mid x_i \in X\}$ released at $A_2 + 4i + 1$ and $A_2 + 4i + 3$. Each dummy job is compatible with all jobs of Q, T^1, F^1, or B_2 and the corresponding opposed dummy job released concurrently.

Proof of Lemma 3.2.2. By Lemmas 3.2.4 and 3.2.3 each rightbound dummy and blocking job of H_2 and B_2 must be scheduled before the end of P_2 and hence, directly at its release. By the given compatibilities this is also true for the leftbound dummy jobs of H_2. Therefore, there are exactly two open slots per variable x_i, one reserved for the two corresponding indefinite jobs $q_i^{\mathrm{t}}, q_i^{\mathrm{f}}$ and one for the two corresponding leftbound variable jobs $f_i^{\mathrm{l}}, t_i^{\mathrm{l}}$. Since no further space is left, for both pairs exactly one can be scheduled within P_2. The remaining one must be completed already by the end of P_1.

Also, for the first part, we can conclude that no blocking and no dummy job released in P_1 can start after the end of P_1. Consider now one variable x_i and assume that no job corresponding to x_i can start within part P_1 after $6i+5$. This assumption holds obviously for x_n. Then, h_i^{rf} and h_i^{lf}, the latest released jobs corresponding to x_i, must both start at their release.

If the leftbound job t_i^{l} is scheduled within part P_1 it must be scheduled at its release and hence $f_{i,1}^{\mathrm{r}}$ and $f_{i,2}^{\mathrm{r}}$ must be postponed to the next parts. In this case, also the second blocking job b_i^{f} as well as the first two dummy jobs h_i^{rt} and h_i^{lt} are forced to start at their release, consequently also b_i^{t}. In this case it is not possible anymore to schedule q_i^{f} within part P_1. For this reason, the counter part q_i^{t} must be scheduled at its release time and the leftbound f_i^{l} must be postponed. With this, there is exactly one free slot for $t_{i,2}^{\mathrm{r}}$ and one for $t_{i,1}^{\mathrm{r}}$.

If, on the other hand, the leftbound job t_i^{l} is scheduled after part P_1, we have to schedule f_i^{l} within part P_1. Due to the conflicts with h_i^{rf}, the start time of f_i^{l} and the blocking and dummy jobs in between must in particular be scheduled at their release. For that reason q_i^{t} must be postponed and q_i^{f} must be scheduled at its release. Hence, also the rightbound true jobs t_i^{r} and t_i^{r} must be postponed and there are exactly two slots for the two false jobs.

In both cases, the scheduled leftbound jobs ensure that no earlier released variable job can start after $6(i-1)+5$. Hence, it can be concluded by induction that, for each variable, either all corresponding false jobs or all corresponding true jobs must be scheduled after part P_1. And since, by Lemmas 3.2.4 and 3.2.3, at least $2n$ rightbound variable jobs must be scheduled within P_1 the free spots ensure that exactly the two counter parts are scheduled within P_1. $\square$

We can conclude the following claim and hence, Theorem 3.2.1.

Claim. There is a satisfying assignment for the given $(\leq 3, 3)$-SAT instance if and only if there is a feasible schedule for the constructed scheduling instance with makespan $C_{\max} = A_5 + 1$.

Proof of Theorem 3.2.1. If there is a schedule with makespan $C_{\max}$ we can apply Lemmas 3.2.4 to 3.2.2. Within the resulting schedule we can therefore be sure that $|C|$ rightbound variable jobs are scheduled within the clause part. Since by Lemma 3.2.2 the assignment of each variable is well defined we get by Lemma 3.2.3 a satisfying truth assignment for the clauses.

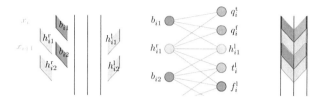

Figure 3.10: Part P_2 creates a structure of blocking and dummy jobs with respective compatibilities that create space for exactly one indefinite job per variable x_i.

If on the other hand a satisfying truth assignment is given, the described schedule with demanded makespan can be created in straight-forward manner, by postponing the assignment jobs corresponding to the truth assignment and scheduling all other jobs within the part they are released in (or in part P_2 in the case of leftbound variable jobs or indefinite jobs). □

3.2.2 Minimization of total completion time

Theorem 3.2.5. *Minimizing the total waiting time for bidirectional scheduling on a single segment with an arbitrary compatibility graph is* NP-*hard even if we assume $p_j = \tau_1 = 1$ for each $j \in J$.*

We give an analogous reduction as for Theorem 3.2.1. Note, that solutions optimal for the total completion time and those optimal for the total waiting time are equivalent. Hence, it is sufficient to prove the hardness for the latter. The goal is to enforce the same structure as for makespan minimization when minimizing the total waiting time. To do so, we start by calculating an upper bound of the resulting waiting time.

We can trivially bound the total waiting time of a schedule that achieves makespan $C_{\max}$ by $W = |J| \cdot C_{\max} = |J| \cdot (A_5 + 1)$, where J is the set of all jobs in our construction. With this polynomial bound we can extend the construction of a scheduling instance for a given ($\leq 3, 3$)-SAT instance by part P_5 with $W + 1$ further leftbound blocking jobs $B_5 = \{b_i \mid i = 0, \ldots, W\}$ with release date $A_5 + i$ for each $b_i^5 \in B_5$ that are not compatible to any of the previous jobs.

Claim. There is a satisfying truth assignment for the given ($\leq 3, 3$)-SAT instance if and only if there is a feasible schedule for the constructed scheduling instance with total waiting time of at most W.

Proof of Theorem 3.2.5. Assume first that there is a satisfying assignment for the ($\leq 3, 3$)-SAT instance. In this case, there is a schedule where no job released in the first four parts starts processing after A_5 and hence the resulting total waiting time does not exceed W.

Assume on the other hand, that there is a solution for the constructed scheduling instance whose objective does not exceed W. For such a solution, either all jobs released in the first four parts start before A_5 or their is at least one starting later. In the first case, we get, by Lemmas 3.2.4 to 3.2.2, a schedule together with a satisfying truth assignment with waiting time bounded by W.

In the second case each postponed job j with starting time S_j' increases the already existing waiting time by at least an amount of $(S_j' - A_5) + W + 1 - (S_j' - A_5) = W + 1$. Hence, the first case applies. □

3.3 Dynamic programs for restricted compatibilities

After establishing the hardness of bidirectional scheduling by exploiting ranges in two different directions, we now consider the case where both properties, the number of segments and the compatibilities, are restricted.

We start with a single segment. If all processing times are equal and no two jobs are compatible the jobs in each direction can simply be scheduled in the order of their release dates. The only decision left is when to switch between left- and rightbound operation of the segments. In the following, we see that this can be tackled by a dynamic program. Such kinds of dynamic programs are also used to solve scheduling problems where setup times are required between jobs of different families, see e.g. the comments of Potts and Kovalyov [PK00].

This idea can even be generalized to the case when some jobs of different directions are compatible, as long as the number of compatibility types is constant. Recall that two jobs have the same compatibility type if they have the same set of compatible jobs on each segment. We partition J into κ subsets of jobs $J^1, \ldots, J^\kappa$ where all jobs of J^c, $c \in 1, \ldots, k$, have the same compatibility type c, and let $n_c = |J^c|$. Since the jobs of each subset only differ in their release dates, they can again be scheduled in the order of their release dates. This allows us to expand the dynamic program to encompass any constant number of compatibility types. We obtain the following result for a single segment.

Theorem 3.3.1. *The total completion time for bidirectional scheduling can be minimized in polynomial time if $m = 1$, κ is constant and $p_j = p$ for each $j \in J$.*

Proof. We consider each subset J^c ordered non-increasingly by release dates and denote by J_i^c the i-th job of J^c in this order, i.e., the $(n_c - i)$-th job to be released. Each entry $T[i_1, t_1, \ldots, i_\kappa, t_\kappa; c]$ of our dynamic programming table is designed to hold the minimum sum of completion times that can be achieved when scheduling only the $i_{c'}$ jobs of largest release date of each compatibility type c', such that $J_{i_{c'}}^{c'}$ is not scheduled before time $t_{c'}$ and $J_{i_c}^c$ is the first job that is scheduled. We start by setting $T[0, t_1, \ldots, 0, t_\kappa; c] = 0$ and define the dependencies between table entries in the following.

Let $C(j, t) = \max\{t, r_j\} + p + \tau_1$ denote the smallest possible completion time of job j when scheduling it not before t. Depending on the types of jobs j_1, j_2 (and in particular of their directions), we can compute in constant time the earliest time $\theta(j_1, t_1, j_2, t_2)$ not before t_1 that job j_1 can be scheduled at, assuming that j_2 is scheduled earlier at time $\max\{t_2, r_{j_2}\}$. We let $\delta_{c'c} = 1$ if $c' = c$ and $\delta_{c'c} = 0$ otherwise, abbreviate $\theta_{c'} = \theta(J_{i_{c'}-\delta_{c'c}}^{c'}, t_{c'}, J_{i_c}^c, t_c)$, and get the following recursive formula for $i_c > 0$:

$$T[i_1, t_1, \ldots, i_\kappa, t_\kappa; c] = \min_{c': i_{c'} \neq 0} \{T[i_1 - \delta_{1c}, \theta_1, \ldots, i_\kappa - \delta_{\kappa c}, \theta_\kappa; c'] + C(J_{i_c}^c, t_c)\}.$$

We can fill out our table in order of increasing sums $\sum i_c$ and finally obtain the desired minimum completion time as $\min_c T[n_1, 0, \ldots, n_\kappa, 0; c]$. We can reconstruct the schedule from the dynamic programming table in straightforward manner. It remains to argue that we only need to consider polynomially many times t_c. This is true, since all relevant times are contained in the set $\{r_j + k\tau + \ell p \mid j, k, \ell \leq n\}$ of cardinality $\mathcal{O}(n^3)$. $\square$

We now consider a constant number of segments $m > 1$. The main complication in this setting is that decisions on one segment can influence decisions on other segments, and, in general, every job can influence every other job in this way. In particular, we need to keep track of how many jobs of each type are in transit at each segment, and we can thus not easily adapt the dynamic program for a single segment. We propose a different dynamic program that relies on all transit times being bounded by a constant.

Theorem 3.3.2. *The total completion time for bidirectional scheduling can be minimized in polynomial time if m, κ, and $\tau_i \in \mathbb{N}$ are constant for each $i \in M$, and $p_j = 1$, $r_j \in \mathbb{N}$ for each $j \in J$.*

Proof. Again, we consider subsets of identical jobs. In addition to their conflict type c, we further distinguish jobs by their start and target segments s, t and form subsets $J^c_{s,t}$ correspondingly. The number of subsets is bounded by κm^2. Since all release times are integer and since $p_j = 1$, we only need to consider integer points in time. Hence, only $\tau_i + 1$ possible positions need to be considered for a job running on segment i, and no two jobs of the same direction can occupy the same position. The state of the system can be fully described by (1) the number of available jobs per segment and $J^c_{s,t}$, and (2) for each position on each segment and each $J^c_{s,t}$, the fact whether a job of $J^c_{s,t}$ is occupying this position. The number of states is bounded by $\prod_{i=1}^{m} n^{\kappa m^2} \cdot \prod_{i=1}^{m} 2^{\kappa m^2 (\tau_i + 1)} = \text{poly}(n)$.

We define the successors of each state to be all states that can be reached in one time step where not all jobs wait, or by waiting for the next release date. This way, the state representation changes from one state to the next. The system always makes progress towards the final state where each job has arrived at its target. The state graph can thus not have a cycle, and we may consider states in a topological order. We formulate a dynamic program that computes for each state the smallest partial completion time to reach the state, where the partial completion time is defined as the sum of completion times of all completed jobs plus the current time for each uncompleted job. The dynamic program is well-defined as each value only depends on predecessor states. $\square$

The exact same idea also works for the case where $p_j = p$, $r_j/p \in \mathbb{N}$ for each $j \in J$ and $\tau_i/p \in \mathbb{N}$ constant. Additionally, we conclude a complementary result to Theorem 3.1.1.

Corollary 3.3.3. *The total completion time for bidirectional scheduling can be minimized in polynomial time if m and κ are constant, $\tau_i = 1$ for each $i \in M$, and $p_j = 0, r_j \in \mathbb{N}$ for each $j \in J$.*

Proof. Since all release dates are integer, at each integer point in time no jobs are running on any segment. We can thus use a simpler version of the dynamic program we introduced in the proof of Theorem 3.3.2. $\square$

4

Competitive Analysis
for Bidirectional Scheduling

We now switch to the online setting where jobs of an instance of bidirectional scheduling are not known in advance but appear by their release date. The purpose of this chapter is to provide theoretical insights on the increase of the cost due to the circumstance that vehicles register their transit request only shortly before their arrival. Compared to the offline setting we show that there is at least a loss of factor 2 even for a single segment. Restricting additionally to identical jobs there is still a loss of at least 1.6. From above, we bound the possible loss for the general BSP by factor 4 based on a respective competitive online algorithm. For special cases, we concretize the algorithm so that it has polynomial running time yielding approximation algorithms. In addition, also the performance ratio can be improved for restricted problem variants.

This chapter is based on joint work with Shashwat Garg and Nicole Megow.

To account for the circumstance that vehicles register their transit request only shortly before their arrival we consider in this chapter an online model for bidirectional scheduling. There, jobs of an instance are not known in advance but appear by their release date. Once, an online algorithm has started a job on a segment it is not possible to revert the decision since it corresponds to the movement of the corresponding vehicle. Transits for later segments can still be adapted, as long as they are not yet started. Hence, if a job on a segment is started it might delay other jobs being released shortly after its start time. This might lead to suboptimal solutions that even can not be prevented with unbounded computational power. We provide an example where we loose independently of the online algorithm's decision at least a factor of $(1+\sqrt{5})/2 > 1.6$ on the offline optimum, even on a single segment with identical jobs, i.e., with unit weights and processing times and all jobs being incompatible (Section 4.2.1). This contrasts to the equivalent special case of single machine scheduling where nothing can be lost by simply scheduling the next available job. For varying processing times we get a lower bound of 2 on possible competitive ratios (Section 4.1.1) which matches the lower bound of 2 for single machine scheduling given by Hoogeveen and Vestjens [HV96].

This lower bound of 2 for total completion time minimization in single machine scheduling is complemented by 2-competitive online algorithms of Hoogeveen and Vestjens [HV96] and Phillips et al. [Phi+95; Phi+98]. In fact, the same competitive ratio is achieved in the weighted case by Anderson and Potts [AP02; AP04]. In addition, Goemans et al. [Goe97; Goe+02] provide online algorithms based on α-points where the randomized variant is 1.69-competitive and the deterministic variant is 2.42-competitive. All these results rely somehow on the analysis of corresponding preemptive relaxations. The unweighted case can be solved by scheduling jobs in SRPT order (shortest remaining processing time first), see [Sch68]. The non-preemptive counterpart schedules jobs in non-decreasing order of their processing time (*SPT order*) which yields an n-competitive algorithm, see [Mao+95]. In the weighted case, scheduling strategies are based on the Smith's ratios which minimizes the mean busy time in the preemptive case.

For bidirectional scheduling, it is not even clear on a single segment how to use these insights about preemptive relaxations. If all jobs are in conflict the first idea that might come to one's mind could be to schedule each subset of aligned jobs $J^d, d \in \{r, l\}$ according to Smith's rule. But still, this gives no insights on decisions when to switch between leftbound and rightbound jobs. And in the presence of compatibilities we need a different model anyhow. By now, there is no reasonable idea of a preemptive relaxation for bidirectional schedules.

Hall et al. [Hal+96; Hal+97] provide a relatively simple but general idea for scheduling problems with weighted completion time objective. For single machine scheduling it yields a 3-competitive online algorithm, i.e., it is not best possible. Nevertheless, due to its universality the presented framework is extremely powerful. Chakrabarti et al. [Cha+96] present polynomial-time variants for other scheduling settings and a randomized version of the algorithm with lower competitive ratio. Krumke et al. [Kru+03] use the idea beyond the scope of machine scheduling to develop a 5.83-competitive online algorithm for the traveling repairman problem. When not asking for polynomial running time the framework also applies to the BSP in its most general variant. This yields a 4-competitive online algorithm for the minimization of total weighted completion time (Section 4.1.2). (We restrict our considerations in this thesis to the analysis of deterministic algorithms.) In the unweighted case we reduce its running time so that it is polynomial for uniform compatibilities on one and two segments. For one segment, the efficient variant can be extended to the weighted case with a slightly increased approximation factor (Section 4.1.3).

Finally, we get back to identical jobs on a single segment. In this special case, a detailed analysis of potential situations is possible. Based on this analysis, we develop a different online algorithm that has a competitive ratio of $(1 + \sqrt{2}) < 2.42$ (Section 4.2.2).

4.1 The general problem

We start by presenting a single segment example that yields a lower bound on any possible competitive ratio. Afterward, we continue by presenting an online algorithm for the general problem which gives an upper bound on the best possible competitive ratio and discuss finally cases allowing for polynomial running time of this algorithm.

4.1.1 Lower bound

Hoogeveen and Vestjens [HV96] provide a lower bound of 2 on the possible competitive ratio in single machine scheduling. Since single machine scheduling is a special case of bidirectional scheduling on a single segment with zero transit times this lower bound directly carries over. For the sake of completeness, we give an adapted version of the proof in [HV96] to show that the 2 also applies if we ask for "a proper bidirectional example" with transit times being not too small and therefore, leading also to an increase of the minimum possible completion time of any job.

Theorem 4.1.1. *Any online algorithm has a competitive ratio of at least 2 even on a single segment.*

Proof. Consider one segment with transit time τ. Assume the adversary releases at time $t = 0$ one job j_1 with processing time p. Any online algorithm A starts j_1 at some point in time $t = S$. If no further job is released by the adversary we have a ratio of $(S + p + \tau)/(p + \tau)$. If the adversary releases k further jobs being opposed and incompatible to j_1 at time $t = S + 1$ with processing time 0 algorithm A has at least a total completion time of $S + p + \tau + k(S + p + 2\tau)$ by scheduling these jobs directly after j_1. The offline optimum can achieve an objective value of at most $k(S + \tau + 1) + (S + p + 2\tau + 1)$ by scheduling the other way around. Figure 4.1 illustrates the two described instances, a best possible answer of A assuming that no more jobs are released, and some schedule Opt. (For the analysis it is sufficient if this schedule yields an upper bound on the optimal objective. Hence, we refrain from proving optimality.) To summarize, we get the following lower bound on the possible competitive ratio:

$$(4.1) \qquad \frac{\mathsf{A}}{\mathsf{Opt}} \geq \max\left\{ \frac{S + p + \tau}{p + \tau}, \frac{(k+1)S + (k+1)p + (2k+1)\tau}{(k+1)(S+1) + p + (k+2)\tau} \right\}.$$

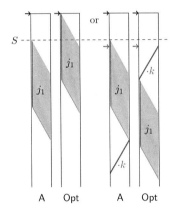
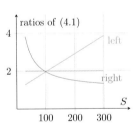

Figure 4.1: Illustration of the two lower bound instances for identical jobs with respective schedules. The corresponding performance ratios are plot in dependence of S with chosen values $k = 150, \tau = 5, p = 100$.

The online algorithm A can choose S to minimize the above value (4.1). This is attained at the intersection point of the two curves defined by the terms in the expression since the left term is monotonically increasing in S while the other is decreasing, cf. the plots of Figure 4.1. Equaling both sides we get:

$$\frac{S + p + \tau}{p + \tau} = \frac{(k+1)S + (k+1)p + (2k+1)\tau}{(k+1)(S+1) + p + (k+2)\tau}$$

and simple transformations yield:

$$1 + \frac{S}{p+\tau} = 1 + \frac{(k-1)\tau - (k+1) + kp}{(k+1)(S+1) + p + (k+2)\tau}.$$

Further transformations give a quadratic equation in S:

$$0 = (k+1)S^2 + (k+1+p+(k+2)\tau)S - (p+\tau)((k-1)\tau + kp - (k+1)).$$

Its positive solution is

$$S_1 = \frac{1}{2(1+k)}\Big(-(k(1+\tau) + 1 + p + 2\tau)$$

$$+ \sqrt{(k+1+p+(2+k)\tau)^2 + 4(1+k)(p+\tau)(-1-k+kp+(k-1)\tau)}\Big)$$

$$= \frac{1}{2(1+k)}\Big(-(k(1+\tau) + 1 + p + 2\tau)$$

$$+ \sqrt{k^2[1 + 4p^2 - 2\tau + 5\tau^2 + p(8\tau - 4)] + \mathcal{O}(k)}\Big)$$

which converges as follows:

$$S_1 \xrightarrow{k \to \infty} \frac{1}{2}\Big(\sqrt{1 + 4p^2 - 2\tau + 5\tau^2 + p(8\tau - 4)} - (\tau + 1)\Big).$$

Therefore, the ratio of (4.1) at the intersection point S_1 tends to

$$\frac{S_1 + p + \tau}{p + \tau} = 1 + \frac{S_1}{p+\tau}$$

$$= 1 + \frac{\sqrt{1 + 4p^2 - 2\tau + 5\tau^2 + p(8\tau - 4)} - (\tau + 1)}{2p + \tau}$$

$$\xrightarrow{p \to \infty} 1 + 1 = 2.$$

$\square$

4.1.2 Upper bound

We use the idea of a general framework for online scheduling of Hall et al. [Hal+96; Hal+97] which can be adapted to our problem in the most general case. Algorithm GR-INT describes the framework. The idea is to consider geometrically increasing time-intervals and to schedule within each interval a subset of jobs available at its beginning such that the weight assigned to this interval is maximal. All jobs released during the interval are not considered before the beginning of the next interval. The choice of each such subset J_x can be implemented with exponential running time by a two-stage enumeration: (1) enumerate over all possible subsets of jobs available at time $t = 2^x$ and (2) enumerate

> **foreach** $x = 1, 2, \ldots$:
> At time $t = 2^x$ choose a subset J_x of the currently unscheduled jobs A_x with maximum total weight $w_x := \sum_{j \in J_x} w_j$ s.t. it admits a schedule starting at 2^x with latest completion time of at most 2^{x+1} and schedule J_x respectively.

Algorithm GR-INT: Greedy online algorithm for bidirectional scheduling operating on exponentially growing intervals.

over all combinatorial frames of J_x and use Algorithm **REALIZE** (Sec. 2.1, page 28) to check for realizability and schedule length.

The analysis of Algorithm **GR-INT** requires implicitly that no job can be completed before time 1. For the sake of completeness we argue in the following why this assumption does not constitute a restriction. Essentially, this works by scaling the time appropriately. The detailed reasoning takes care that this scaling can be done respecting the online setting, i.e., before assigning the first start time the scaling factor must be assessed without knowing all jobs.

Lemma 4.1.2. *W.l.o.g. we can assume for Algorithm* **GR-INT** *that* $r_j + \sum_{i \in M_j} (p_j + \tau_i) \geq 1$ *for each job* $j \in J$.

Proof. We first define for each job $j \in J$:

$$
q_j := \begin{cases} r_j & \text{if } r_j > 0 \\ p_j & \text{if } r_j = 0, p_j > 0 \\ \tau_{s_j} & \text{otherwise.} \end{cases}
$$

First, we can assume that each $q_j > 0$ since otherwise we can start j on segment s_j at time $t = 0$ without any delay for j or any other job.

Now consider $q := \min_{j \in J} q_j$. If $q \geq 1$ the claim follows. Otherwise, no job has been started by Algorithm **GR-INT** before time q. Afterward, Algorithm **GR-INT** can work with a stretched time-scale, i.e., with adapted time values $r'_j := r_j/q$, $p'_j := p_j/q$ for each $j \in J$, and $\tau'_i := \tau_i/q$ for each $i \in M$. Each job j is then actually scheduled at time $S_j := qS'_j$ instead of the start time S'_j assigned by **GR-INT**. Note that scaled optimal offline solutions coincide, i.e., S^*_j/q defines an optimal solution for the scaled version and vice versa. Consequently, an achieved performance ratio for the scaled instance also applies for the original instance. $\square$

Let **Opt** be an optimal schedule and let it also denote its objective value. Denote the completion times scheduled by **GR-INT** as C_j, those of **Opt** as C^*_j, and the x-th interval by $I_x := (2^x, 2^{x+1}]$ for $x > 0$ and $I_0 := [2^0, 2^1]$. Let $J^*_x := \{j \in J \mid C^*_j \in I_x\}$ with total weight $w^*_x := \sum_{j \in J^*_x} w_j$ be the set of jobs scheduled by **Opt** to complete in interval I_x. Due to Lemma 4.1.2 we can assume that no job is completed before I_0, i.e., we know that $J = \bigcup_{x \geq 0} J^*_x$. Hence, we can analyze the performance of **GR-INT** as follows.

Theorem 4.1.3. *Algorithm* **GR-INT** *is* 4-*competitive.*

Proof. Since **GR-INT** does not assign jobs before the beginning of I_1 we know that $w_0 = 0$ while w^*_0 can be positive. Due to the defined interval length of $|I_\ell| = 2^{\ell+1} - 2^\ell = 2^\ell$ all jobs $j \in \cup_{x < \ell} J^*_x$ scheduled by **Opt** fit in the interval I_ℓ. In particular, all jobs from $\cup_{x < \ell} J^*_x$

that have not been scheduled by GR-INTERVAL by time 2^ℓ, are available at time 2^ℓ and they fit in I_ℓ. Since GR-INTERVAL maximizes the weight of the chosen jobs, the actual weight assigned to the first ℓ intervals is as least as high as the total weight of jobs being completed by Opt before 2^ℓ, and hence,

$$(4.2) \qquad \sum_{x=1}^{\ell} w^*_{x-1} \leq \sum_{x=1}^{\ell} w_x \qquad \text{for all } 1 \leq \ell.$$

This holds in particular for the smallest index L with $|J^*_\ell| = 0$ for all $\ell \geq L$. Therefore and since no job is released after 2^L we get that no job scheduled by GR-INT completes later than 2^{L+1}, yielding

$$(4.3) \qquad \sum_{x=1}^{L} w^*_{x-1} = \sum_{x=1}^{L} w_x.$$

By definition we know that any job in J_x is completed by GR-INT not later than 2^{x+1} and any job in J^*_x is completed by Opt after 2^x. Hence, using (4.2) and (4.3) we can apply Lemma 4.1.4 (see below) assuming that w^*_{x-1} corresponds to a_i, w_x to b_i and 2^{x+1} to R_i. Therefore, we get:

$$\sum_{j \in J} w_j C_j = \sum_{x=1}^{L} \sum_{j \in J_x} w_j C_j \leq \sum_{x=1}^{L} 2^{x+1} w_x$$

$$\leq \sum_{x=1}^{L} 2^{x+1} w^*_{x-1} = 4 \sum_{x=0}^{L-1} 2^x w^*_x \leq 4\mathsf{Opt}.$$

$\square$

Within the proof of Theorem 4.1.3 we apply the following fact that can be proven via induction, cf. [Kru+03].

Lemma 4.1.4. *Let $a_i, b_i, i = 1, \ldots, p$ be non-negative values such that:*
(a) $\sum_{i=1}^{p'} a_i \leq \sum_{i=1}^{p'} b_i$ for all $1 \leq p' \leq p$ and
(b) $\sum_{i=1}^{p} a_i = \sum_{i=1}^{p} b_i$.
Then $\sum_{i=1}^{p} R_i b_i \leq \sum_{i=1}^{p} R_i a_i$ for any non-decreasing sequence $0 \leq R_1 \leq \ldots \leq R_p$.

Proof. The base case for $p = 1$ holds obviously since $a_1 = b_1$ implies $R_1 a_1 \leq R_1 b_1$ for any non-negative value R_1. Assume for the induction step that the statement holds for some integer $p \geq 1$. Consider now values a_i, b_i and $R_i, i = 1, \ldots, p + 1$ such that $0 \leq R_1 \leq \ldots \leq R_p \leq R_{p+1}$, $\sum_{i=1}^{p+1} a_i = \sum_{i=1}^{p+1} b_i$, and $\sum_{i=1}^{p'} a_i \leq \sum_{i=1}^{p'} b_i$ for all $1 \leq p' \leq p + 1$. The combination includes in particular that $0 \leq \sum_{i=1}^{p} b_i - \sum_{i=1}^{p} a_i = a_{p+1} - b_{p+1}$.
We consider now the adapted sequence $a'_i = a_i, b'_i = b_i, i = 1, \ldots, p - 1$ and $a'_p = a_p + a_{p+1}, b'_p = b_p + b_{p+1}$ satisfying the conditions of the induction hypothesis due to $\sum_{i=1}^{p} a'_i = \sum_{i=1}^{p+1} a_i = \sum_{i=1}^{p+1} b_i = \sum_{i=1}^{p} b'_i$. Therefore, we can apply the induction hypothesis and get

with $R_{p+1} \geq R_p$:

$$\sum_{i=1}^{p+1} R_i a_i - \sum_{i=1}^{p+1} R_i b_i = \sum_{i=1}^{p} R_i a_i - \sum_{i=1}^{p} R_i b_i + \underbrace{(a_{p+1} - b_{p+1})}_{\geq 0} R_{p+1}$$

$$\geq \sum_{i=1}^{p} R_i a_i' - \sum_{i=1}^{p} R_i b_i' \geq 0.$$

$\square$

4.1.3 Polynomial running time

Algorithm GR-INT gives an upper bound on the ratio that is lost by the online arrivals of the vehicles. It remains the question of its time complexity. For some special cases of one and two segments with uniform compatibilities we provide polynomial time variants for the respective subroutines.

Single segment

We start with uniform compatibilities on a single segment. If all opposed jobs have a conflict, a schedule of minimum length can be determined straight forward by scheduling first all jobs of one and afterward all jobs of the other direction. If all jobs are compatible it gets even simpler. If considering only one segment we denote the transit time of this segment shortly by $\tau := \tau_1$.

Lemma 4.1.5. *The schedule length of a set of available jobs J can be minimized in linear time if a single segment with uniform compatibilities is considered.*

Proof. Two non-empty sets of rightbound J^r and leftbound J^l jobs can be scheduled with length

$$(4.4) \qquad L(J^r, J^l) = \begin{cases} p(J^r) + p(J^l) + 2\tau & \text{if } G_1 = \emptyset \\ \max\{p(J^r), p(J^l)\} + \tau & \text{if } G_1 = K_{n_r, n_l} \end{cases}$$

by either scheduling first all jobs of J^r in arbitrary order and afterward all jobs of J^l in arbitrary order if they have conflicts or scheduling a sequence of J^r concurrently with a sequence of J^l if they are compatible. Note that $L(J^r, J^l)$ is also a lower bound for the scheduling length of J since no two aligned jobs can be processed concurrently. Hence, for some set J a schedule of minimum length can be constructed in polynomial time. Similar (even simpler) observations hold if one of the both sets is empty. $\square$

We now specify how to choose the subset to be scheduled within every interval. In the unweighted case, we only need to maximize the number of assigned jobs. For this setting, we can restrict per heading to a fixed order to choose in. Consequently, the number of possibilities reduces to be quadratic.

Lemma 4.1.6. *The choice of J_x in GR-INT can be executed in $\mathcal{O}(n^2)$ if a single segment with uniform compatibilities and equal weights is considered.*

Proof. Consider the right- and leftbound jobs A_x^{r} and A_x^{l} available and unscheduled by algorithm GR-INT at time $t = 2^x$. By a simple exchange argument we can choose the jobs J_x^d of each direction $d \in \{\mathrm{r}, \mathrm{l}\}$ to be scheduled in I_x without loss in SPT order: Assume there are jobs $j_1 \in A_x^d \setminus J_x^d$ and $j_2 \in J_x^d$ with $p_{j_1} < p_{j_2}$. Job j_2 can be feasibly replaced by j_1 without changing any other job and hence, without increasing the schedule length. Therefore, we let J_x^d for each direction $d \in \{\mathrm{r}, \mathrm{l}\}$ to be the first n_d jobs of A_x^d in SPT-order, try all possible combinations of $(n_{\mathrm{r}}, n_{\mathrm{l}})$, and choose one that yields the maximum number of jobs with resulting length of at most 2^x. The number of possible combinations is quadratic in the number of jobs. □

With these two ingredients we get a quadratic running time for each iteration and can therefore conclude the following.

Theorem 4.1.7. *Algorithm GR-INT is a 4-approximation algorithm for total completion time minimization on a single segment with uniform compatibilities.*

Proof. First observe that the number of intervals to be considered is bounded by $\log U$ where $U = \max_j r_j + n \cdot (\max_j p_j + \tau)$ is some upper bound on the necessary makespan. Hence, by Lemmas 4.1.5 and 4.1.6 we get a running time of $\mathcal{O}(n^2)$. The approximation factor follows by Theorem 4.1.3. □

With Lemma 4.1.5 we are also able to extend the idea of Hall et al. [Hal+97] for the weighted case. To that end, the technique of ρ-*dual approximation algorithms* is used, confer Hochbaum and Shmoys [HS87]. The key observation is that finding a subset of jobs with total length not exceeding a given bound but having maximum weight is equivalent to the knapsack problem, see the comprehensive book of Kellerer et al. [Kel+04].

Knapsack Problem (KP)

Given: A set of n items with size p_j and value w_j for each $j = 1, \ldots, n$, a capacity C.

Task: Find a subset S^* of items with maximum value $w(S^*) := \sum_{j \in S^*} w_j$ such that $p(S^*) := \sum_{j \in S^*} p_j \leq C$.

The KP admits dynamic programs with pseudo-polynomial running time yielding fully polynomial time approximation schemes, confer Ibarra and Kim [IK75] for the earliest result and Kellerer et al. [Kel+04] for detailed discussions. Dynamic programming additionally allows for $(1+\varepsilon)$-dual approximation schemes. A ρ-dual approximation algorithm for the KP finds a subset of items with a total value $w' \geq w(S^*)$ such that its total size is at most ρC. For our construction we assume that the KP admits for every $\varepsilon > 0$ a $(1+\varepsilon)$-dual approximation algorithm with running time in $\mathcal{O}(n^2 \frac{1}{\varepsilon})$ and refer for further details to [Hal+97].

Lemma 4.1.8. *Consider a set of available jobs A with uniform compatibilities on a single segment. Let w be the maximum amount of weight that can be scheduled with a length of at most L. For every $\varepsilon > 0$ we can find in $\mathcal{O}(n^2 \frac{1}{\varepsilon})$ a subset of A admitting a schedule of length $(1 + \varepsilon)L$ that has a total weight of $w' \geq w$.*

Proof. If all opposed jobs are compatible, both headings can be considered independently. The stated result can be achieved by applying for each $d \in \{r, l\}$ a knapsack $(1 + \varepsilon)$-dual approximation on the set A^d with weights as values and processing times as sizes and the capacity $L - \tau$. Then, the resulting schedule has a length of at most $(1 + \varepsilon)(L - \tau) + \tau \leq (1 + \varepsilon)L$ and a total weight $w' \geq w$ due to (4.4) of Lemma 4.1.5.

If no two jobs are compatible we take the subset with maximum weight out of the following three approximate knapsack solutions with processing times as sizes and weights as values: (1) items A^r with capacity $L - \tau$, (2) items A^l with capacity $L - \tau$, or (3) items A with capacity $L - 2\tau$. In any of the three cases we get a resulting schedule with a length of at most $(1 + \varepsilon)L$ due to (4.4) of Lemma 4.1.5. Since each optimal schedule corresponds to one of the three possibilities we conclude for the chosen subset that $w' \geq w$. $\square$

Hence, we can adapt Algorithm GR-INT with a small increase of the objective value for the benefit of a polynomial running time. To that end, we denote for every $x \geq 1$ the interval $((1 + \varepsilon)2^x, (1 + \varepsilon)2^{x+1}]$ by I'_x. Now, Algorithm GR-INT$(1 + \varepsilon)$ considers at time 2^x the set A'_x of unscheduled jobs released up to time 2^x, chooses a subset J'_x of A'_x according to Lemma 4.1.8 for length 2^x and schedules J'_x according to Lemma 4.1.5 within I'_x that has a length of $(1 + \varepsilon)2^{x+1} - (1 + \varepsilon)2^x = (1 + \varepsilon)2^x$.

Theorem 4.1.9. *Algorithm GR-INT$(1+\varepsilon)$ is a $4(1+\varepsilon)$-approximation algorithm for total weighted completion time minimization on a single segment with uniform compatibilities.*

Proof. Lemma 4.1.8 yields a running time of $\mathcal{O}(n^2 \frac{1}{\varepsilon})$ as in the proof of Theorem 4.1.7 and the respective inequality corresponding to (4.2) for the total weights w'_x of each set J'_x. Similar calculations as in the proof of Theorem 4.1.3 yield the result since each job $j \in J'_x$ is completed before time $(1 + \varepsilon)2^{x+1}$.

$$\sum_{j \in J} w_j C_j = \sum_{x=1}^{L} \sum_{j \in J'_x} w_j C_j \leq \sum_{x=1}^{L} (1 + \varepsilon)2^{x+1} w'_x$$

$$\leq \sum_{x=1}^{L} (1 + \varepsilon)2^{x+1} w^*_{x-1} = (1 + \varepsilon)4 \sum_{x=0}^{L-1} 2^x w^*_x \leq (1 + \varepsilon)4\mathsf{Opt}.$$

$\square$

In single machine scheduling it is actually possible to improve the framework such that it is 3-competitive, cf. [Hal+97]. The idea is to schedule the set of chosen jobs J_x within I_x such that the sum of (weighted) completion times is minimized. This is possible since a sequence of jobs on a single machine can be rearranged without increasing its length. Unfortunately, this property does not hold if bidirectional jobs have conflicts. Figure 4.2 presents a short example where the schedule described above (proof of Lemma 4.1.5) is not optimal wrt. total completion time.

In the case of all opposed jobs being compatible an arbitrary rearrangement of each side is possible without increasing the length of the schedule. For the sake of completeness, we state the following result and refer for the detailed analysis to [Hal+97].

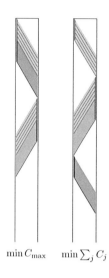

Figure 4.2: Example comparing minimum makespan and minimum total completion time. For each direction $d \in \{r, l\}$ we release at time $t = 0$ a set of k jobs with processing time p_1 and one job with processing time $p_2 > p_1 + 2\tau/k$. The above minimum length schedule has the following sums of completion times for (i) the short rightbound jobs: $k(k+1)/2 \cdot p_1 + k\tau$, (ii) the following long rightbound job: $kp_1 + p_2 + \tau$, (iii) the short leftbound jobs: $k(kp_1 + p_2 + 2\tau) + k(k + 1)/2 \cdot p_1$, and (iv) the last long job: $2n_1p_1 + 2p_2 + 2\tau$. In total we get

$$(2n_1^2 + 3n_1)p_1 + (k + 3)p_2 + (3n_1 + 3)\tau.$$

Moreover, we can schedule the jobs with decreased total completion time: (i) first all short rightbound jobs: $k(k + 1)/2 \cdot p_1 + k\tau$, (ii) then all short leftbound jobs: $k(kp_1 + 2\tau) + k(k + 1)/2 \cdot p_1$, (iii) followed by the long leftbound job: $2n_1p_1 + p_2 + 2\tau$, and (iv) finally the long rightbound job: $2n_1p_1 + 2p_2 + 3\tau$ which is in total:

$\min C_{\max}$ $\min \sum_j C_j$ $(2n_1^2 + 4n_1)p_1 + 3p_2 + (3n_1 + 5)\tau.$

Hence, there is a difference of $kp_2 - kp_1 - 2\tau > 0$.

Corollary 4.1.10. *Minimizing the total (weighted) completion time on a single segment admits a polynomial time online algorithm with approximation factor 3 (resp. $3+\varepsilon$) if any two opposed jobs are compatible.*

Two segments

We continue now with two segments. Also for this setting and uniform compatibilities we can create a simple schedule with minimum length. The rough idea is to send first all jobs from the boundaries to the central node between the two segments and afterward all jobs from the central node to the boundaries. If this is done in an appropriate way there is only one switch of direction between two consecutive jobs per segment and one can show that the resulting schedule length is tight to either the amount of work of the corresponding segment or the minimum possible transit time of those jobs that need to cross both segments.

Lemma 4.1.11. *The schedule length of a set of available jobs J can be minimized in linear time if two segments with uniform compatibilities are considered.*

Proof. The given set of jobs J can be partitioned into the subsets $J^{r,s,t}, 1 \leq s \leq t \leq 2$ and $J^{l,s,t}, 1 \leq t \leq s \leq 2$ defined by source and target segment of each job as well as its heading. Denote the number of demanded headings per segment i as ℓ_i.

There are two types of lower bounds on the schedule length. First, the minimum amount of processing and transit for each segment $i = 1, 2$:

$$L^i = \begin{cases} p(J^{r,i,i}) + p(J^{r,1,2}) + p(J^{l,2,1}) + p(J^{l,i,i}) + \ell_i\tau_i & \text{if } G_i = \emptyset \\ \max\{p(J^{r,i,i}) + p(J^{r,1,2}), p(J^{l,2,1}) + p(J^{l,i,i})\} + \tau_i & \text{if } G_i = K_{n_r,n_l}. \end{cases}$$

Second, we consider jobs that need to transit both segments. We prove for each set $J^{d,s,t}$ with $(d,s,t) \in \{(\mathrm{r},1,2),(\mathrm{l},2,1)\}$ the following lower bound on the schedule length:

$$L^d = p_{\max} + \tau_s + p(J^{d,s,t}) + \tau_t$$

where $p_{\max} = \max_{j \in J^{d,s,t}} p_j$. Consider some job $j \in J^{d,s,t}$ with processing time $p_{\max}$ and some schedule for $J^{d,s,t}$. For segment i let S_{ij} be the start time of j on i and P_i be the subset of jobs being scheduled before j on i. We partition P_t into P_t^1 and P_t^2 such that $P_t^1 \subseteq P_s$ and $P_t^2 \cap P_s = \emptyset$. Let j' be the first job of P_t^2 scheduled on t with start time $S_{tj'}$. Since $S_{sj} \geq p(P_s) \geq p(P_t^1)$ and j' is scheduled after j on s, we get $S_{tj'} \geq S_{sj} + p_{\max} + \tau_s \geq p(P_t^1) + p_{\max} + \tau_s$. Hence, we get $S_{tj} \geq S_{tj'} + p(P_t^2) \geq p(P_t) + p_{\max} + \tau_s$. (If P_t^2 is empty, this inequality follows directly.) Since j is processed on t after its start time S_{tj} followed by the processing of its succeeding jobs and the last transit τ_t, we get the above lower bound.

We now construct the following schedule by defining a partial order of each segment. We then let each job start as early as possible after each incompatible predecessor in this order and possibly its completion time on the preceding segment. On segment 1 we schedule first all jobs of $J^{\mathrm{r},1,2}$ in non-increasing order of their processing times, afterward all jobs of $J^{\mathrm{r},1,1}$. Respectively, on segment 2 we schedule first all jobs of $J^{\mathrm{l},2,1}$ in non-increasing order of their processing times and afterward all jobs of $J^{\mathrm{l},2,2}$. We continue on segment 1 with all jobs of $J^{\mathrm{l},1,1}$, followed by all jobs of $J^{\mathrm{l},2,1}$ in the same order. Similarly, on segment 2 we continue with all jobs of $J^{\mathrm{r},2,2}$ followed by all jobs of $J^{\mathrm{r},1,2}$ in the same order. Let t be the segment with latest completion time. If each job starts subsequently after its direct predecessor on t the schedule length is equal to L^t and hence the schedule is optimal. Otherwise, some job j on segment t with direction d has to wait for its completion on the previous segment s. Due to the given orders this can only be the first scheduled job of $J^{d,s,t}$ with longest processing time. Hence, the corresponding schedule length equals L^d. □

Similar considerations as for the single segment yield that also the choice of jobs to be scheduled within each iteration can be implemented in polynomial time if we only need to maximize the number of assigned jobs.

Lemma 4.1.12. *The choice of J_x in GR-INT can be executed in $\mathcal{O}(n^6)$ if two segments with uniform compatibilities and equal weights are considered.*

Proof. As in the previous Lemma there are 3 relevant rightbound and 3 relevant leftbound subsets of the available jobs with invariant heading, start and target segment. By the same exchange argument as in Lemma 4.1.6 we can choose the jobs from each such subset to be scheduled in I_x without cost increase in SPT order. Assume there is a scheduled job j_1 and an unassigned job j_2 with $p_{j_2} < p_{j_1}$. Job j_1 can be feasibly replaced by j_2 without changing any other job and hence, without increasing the schedule length. There are $\mathcal{O}(n^6)$ possible combinations of possible SPT subsets from the above six job sets. It is sufficient to test for each combination if it admits a schedule of length at most 2^x to get the choice with maximum cardinality. □

With these two ingredients we get a polynomial running time for each iteration and can therefore conclude the following theorem with a similar proof as for Theorem 4.1.7.

Theorem 4.1.13. *Algorithm GR-INT is a 4-approximation algorithm for total completion time minimization on two segments with uniform compatibilities.*

4.2 Identical jobs on a single segment

We consider now the case that all jobs have a conflict and equal processing time p on a single segment with transit time τ. In the offline setting there is an exact algorithm solving any such instance in polynomial time, see Section 3.3. However, the presented dynamic program needs the complete instance in advance. If jobs arrive online such a dynamic program does not work out. In the following section we present an example that it is not possible to achieve the offline optimum in the online setting. This contrasts to the equivalent special case of single machine scheduling where nothing is lost by simply scheduling the next available job.

4.2.1 Lower bound

Theorem 4.2.1. *Any online algorithm has a competitive case ratio of at least $(1 + \sqrt{5})/2 \approx 1.618$ even on a single segment with identical jobs.*

Proof. Assume the adversary releases at time $t = 0$ one job j_1 with processing time $p = 0$. Any online algorithm A has to schedule j_1 to start at some point $t = S$ in time. If no further job is released by the adversary we have a ratio of $(S + \tau)/\tau$. If the adversary releases k further jobs opposed to j_1 at time $t = S + 1$ algorithm A has at least a total completion time of $S + \tau + k(S + 2\tau)$ whereas the offline optimum can achieve an objective value of $k(S + \tau + 1) + (S + 2\tau + 1)$ or better. Figure 4.3 illustrates the two described instances, a best possible answer of A assuming that no more jobs are released, and some schedule for an upper bound of Opt. To summarize, we get the following lower bound on the possible competitive ratio:

$$(4.5) \qquad \frac{\mathsf{A}}{\mathsf{Opt}} \geq \max\left\{ \frac{S + \tau}{\tau}, \frac{(k + 1)S + (2k + 1)\tau}{(k + 1)(S + 1) + (k + 2)\tau} \right\}.$$

The online algorithm A can choose S to minimize the above value (4.5). This is attained at the intersection point of the two curves representing the terms in the expression since the left term is monotonically increasing in S while the other is decreasing, cf. the plots of Figure 4.3. Equaling both sides we get:

$$\frac{S + \tau}{\tau} = \frac{(k + 1)S + (2k + 1)\tau}{(k + 1)(S + 1) + (k + 2)\tau}$$

and simple transformations yield:

$$1 + \frac{S}{\tau} = 1 + \frac{(k - 1)\tau - (k + 1)}{(k + 1)(S + 1) + (k + 2)\tau}.$$

We now let $\tau = k$ and get by further transformations the following quadratic equation in S

$$0 = (k + 1)S^2 + (k^2 + 3k + 1)S - (k^3 - 2k^2 - k).$$

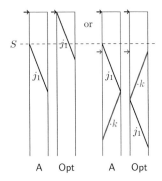

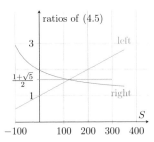

Figure 4.3: Illustration of the two lower bound instances for identical jobs with respective schedules. The corresponding performance ratios are plot in dependence of S with chosen values $\tau = k = 200$.

Dividing the non-negative of the both roots by k we get

$$\frac{S}{k} = \frac{-(k^2 + 3k + 1) + \sqrt{5k^4 + 2k^3 - k^2 + 2k + 1}}{2k(k + 1)}.$$

With $k \to \infty$ the right hand side converges to $\frac{-1+\sqrt{5}}{2}$. Consequently, $1 + \frac{S}{k}$ and therefore the intersection point tends to $\frac{1+\sqrt{5}}{2}$. $\qquad\square$

4.2.2 Upper bound

Consider some schedule S. For each job $j \in J$ we denote by $\ell_j(S)$ the number of jobs completed by time C_j in schedule S (including j). Since $\{\ell_j(S) \mid j \in J\} = \{1, \ldots, n\}$ it holds that $\sum_{j \in J} \ell_j(S) = \frac{n(n+1)}{2}$. Since each job j contributes at least $\ell_j(S)p + \tau$ to the objective value of S and this holds in particular for Opt we can conclude that

$$(4.6) \qquad \sum_{j \in J}[\ell_j(S)p + \tau] = \frac{n(n+1)}{2}p + n\tau \le \mathsf{Opt}$$

and get a lower bound on the optimum. We also use a second obvious lower bound:

$$(4.7) \qquad \sum_{j \in J}[r_j + p + \tau] \le \mathsf{Opt}.$$

We now present an online algorithm for the considered special case that is motivated by the following observation.

Observation 4.2.2. *If at the end of processing of some job j ($t = S_j + p$) an aligned job is available it can be scheduled at time t without any loss.*

When adapting some schedule job by job to satisfy the above property no additional transit time is inserted between two processing intervals and the increased completion time by value p per delayed job is compensated by the saved completion time of the antedated job. Or in other words, the values of ℓ_j are rearranged but still sum up from 1 to n and no completion time is increased by a transit τ.

---

input: Jobs J arriving online, preferred start direction $d \in \{\mathrm{r}, \mathrm{l}\}$
Let $t := 0$
while there are unscheduled jobs:
    if $t = S_j + p$ of some running job $j \in J^d$:
        if some job $k \in J^d$ is available at time t:
            Start k at time $S_k = t$ and increase t by p
        else:
            Let d be the opposite and increase t by τ
    elif some job $k \in J^d$ is available at time t:
        Start k at time $S_k = t$ and increase t by p
    else:
        if a job $k \in J \setminus J^d$ is available:
            Start k at time $S_k = t$, increase t by p, and let d be the opposite
        else:
            Let t be the next release date and d the corresponding direction.

Algorithm SWITCH: As long as aligned jobs of the currently processed job j are available at time $S_j + p$ continue with these. Otherwise switch the direction.

Therefore, we define Algorithm **SWITCH** that obeys the described property. If a job finishes processing and a further aligned job is available it is scheduled. On the other hand, if a job is released while each started aligned job has finished processing but some is still in transit we do not start the released job even though it would be feasible. For this reason we are able to bound the time between the release of a job and its completion as follows:

Lemma 4.2.3. *Consider a schedule created by* **SWITCH**. *For any job* $j \in J$ *it holds that*

$$(4.8) \qquad\qquad C_j \leq r_j + 3\tau + \ell_j p.$$

Proof. Consider some job j. Note that **SWITCH** does not create idle time as long as a job is available. Hence, at each point in time between release and completion of j some job is processed or in transit. The last such job is j itself yielding one τ of the upper bound. The processing intervals of all jobs completing by time C_j are covered by $\ell_j p$. We only have to bound the remaining intervals covered by transit of some job but not by processing.

If j is released while an aligned job is processed j is started without an additional transit interval. If j is released while an opposed job is running it is started the next time the direction is switched. Hence, one additional transit interval is needed. This holds in particular because the direction switch is applied immediately by the time no further opposed job is available. Either of these two cases applies if no job is running when j is released since at time r_j some aligned or opposed job is started. The third transit interval appears if j is released after the processing of each scheduled aligned job and before opposed jobs could have been started. $\qquad\square$

Combining Lemma 4.2.3 with the above lower bounds (4.6) and (4.7) we can proof that SWITCH is 3-competitive. With the tight example presented in Figure 4.4 we even can conclude that it has a competitive ratio of 3.

Figure 4.4: Tight Example for SWITCH: At time $t = 0$ we release one rightbound job and at time ε we release one leftbound job as well as k rightbound jobs, all having processing time $p = 0$. Algorithm SWITCH starts with the first rightbound job, followed by the leftbound job, and finally the k rightbound jobs. This yields a total completion time of $3\tau + k \cdot 3\tau$. Scheduling first all rightbound jobs at their release and afterward the leftbound job yields an objective value of $3\tau + \varepsilon + k \cdot (\tau + \varepsilon)$. Hence, the ratio of SWITCH and Opt converges to $\frac{3\tau}{\tau + \varepsilon}$ as k tends to ∞ which is arbitrarily close to 3.

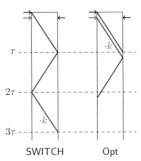

For jobs being released very early also the lower bound that depends on the release date is small. A common technique to deal with these jobs is to wait for a certain amount of time such that sufficient knowledge is available how to schedule all early jobs, cf. e.g. [HV96]. If a good tradeoff between delay and gained knowledge can be found we can decrease the proven performance guarantee. Algorithm D-SWITCH describes the delayed variant of Algorithm SWITCH.

---

input: Jobs J arriving online, parameter Δ for the delay
Define new release dates $r'_j = max\{r_j, \Delta\}$
Let d be the direction with more jobs released before Δ
Run SWITCH for modified release dates with first preference to d

---

Algorithm D-SWITCH: Delayed SWITCH algorithm.

We now discuss two variants depending on the chosen parameter Δ. To do so, we partition the set of jobs into $J_1 := \{j \in J \mid r_j < \Delta\}$ and $J_2 := \{j \in J \mid r_j \geq \Delta\}$ with cardinalities n_1 and n_2. We denote the completion times of Opt(J) by C_j^* and those of D-SWITCH(J) by C_j and use the respective abbreviations $\ell_j^* := \ell_j(\text{Opt}(J))$ and $\ell_j := \ell_j(\text{D-SWITCH}(J))$. Let furthermore $\ell_0^* \leq n_1$ be the number of jobs started by Opt before time τ. We finally define

$$P_1^* := \sum_{j \in J_1} \ell_j^* p \text{ and } P_2^* := \sum_{j \in J_2} \ell_j^* p \quad \text{as well as} \quad P_1 := \sum_{j \in J_1} \ell_j p \text{ and } P_2 := \sum_{j \in J_2} \ell_j p.$$

Lemma 4.2.4. *For any schedule created by D-SWITCH with $\Delta = \sqrt{2}\tau$ it holds that*

$$\sum_{j \in J} C_j \leq (1 + \sqrt{2})\text{Opt} \leq 2.415\text{Opt}.$$

Proof. Consider the relaxation of J_1 where all jobs are released at time $t = 0$. Let $d \in \{r, l\}$ be the direction with more jobs in J_1. Note that it is optimal to schedule first all jobs of direction d and afterward the remaining. Denote the respective completion times by C'_j.

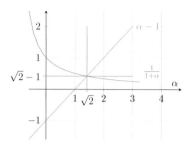

Figure 4.5: Plots of $\alpha - 1$ and $\frac{1}{1+\alpha}$ with intersection point $(\sqrt{2}, \sqrt{2} - 1)$. It yields the minimum possible value of $\max\{\alpha - 1, \frac{1}{1+\alpha}\}$ for non-negative α.

Algorithm **SWITCH** is asked to prefer direction d but start by time Δ. There might be jobs of J_2 scheduled before some job $j \in J_1$. The respective delay is bounded by $\ell_j^* p$. Hence, we can bound the costs of J_1 as follows:

$$(4.9) \qquad \sum_{j \in J_1} C_j \leq \sum_{j \in J_1} C_j' + n_1 \Delta + P_1.$$

The completion times of J_2 are again bounded by Lemma 4.2.3. Let $\alpha \geq 1$ such that $\Delta = \alpha \tau$. Since no job of J_2 is released before time Δ we get the lower bound $(1 + \alpha)\tau \leq C_j^*$ for each $j \in J_2$. Applying additionally (4.6) and (4.7) we get

$$\sum_{j \in J} C_j \leq \sum_{j \in J_1} C_j' + n_1 \Delta + P_1 + \sum_{j \in J_2} (r_j + 3\tau + \ell_j p)$$

$$= \sum_{j \in J_1} C_j' + \sum_{j \in J_2} (r_j + \tau + p) + \alpha \tau n_1 + 2\tau n_2 + P_1 + P_2$$

$$\leq \sum_{j \in J} C_j^* + \frac{n(n+1)}{2} p + \tau n + (\alpha - 1)\tau n_1 + \tau n_2$$

$$\leq 2 \sum_{j \in J} C_j^* + (\alpha - 1) \sum_{j \in J_1} C_j^* + \frac{1}{1+\alpha} \sum_{j \in J_2} C_j^*$$

$$= \left(2 + \max\{\alpha - 1, \frac{1}{1+\alpha}\} \right) \sum_{j \in J} C_j^*.$$

Since $\alpha - 1$ is monotonically increasing and $\frac{1}{1+\alpha}$ is monotonically decreasing on the relevant domain, the smallest possible factor of the above evaluation is attained at their point of intersection being $(\sqrt{2}, \sqrt{2} - 1)$ since $\alpha - 1 = 1/(1 + \alpha)$ is equivalent to $\alpha^2 = 2$, cf. Figure 4.5. □

Considering the choice $\Delta = \tau$ indicates that an improved performance ratio by a more detailed analysis might be possible. The following lemma summarizes the currently best version yielding a factor of 2 for special cases.

Lemma 4.2.5. *For any schedule created by* D-SWITCH *with* $\Delta = \tau$ *it holds that*

$$\sum_{j \in J} C_j \leq 2\mathsf{Opt} + n_2 \min\{\ell_0^* p, \tau\} - n_1^2 p \leq 2.5\mathsf{Opt}.$$

This yields in particular $\sum_{j \in J} C_j \leq 2\mathsf{Opt}$ *if* $p = 0$, $n_1 = 0$ *or* $n_1 \geq n_2$.

Proof. For the jobs of J_1 we argue as for Lemma 4.2.4 but bound the delay for $j \in J_1$ by $(\ell_j^* - n_1)p$ since we need to respect only the jobs of J_2.

$$\sum_{j \in J_1} C_j \leq \sum_{j \in J_1} C_j' + \tau n_1 + P_1^* + P_1 - P_1^* - n_1^2 p$$

$$\leq 2 \sum_{j \in J_1} C_j^* + P_1 - P_1^* - n_1^2 p.$$

For each job $j \in J_2$ we can give an additional bound of $C_j^* \geq 2\tau + \ell_j^* p - \min\{\ell_0^* p, \tau\}$ respecting the processed jobs between j's release date and its start which is in many cases stronger. Combining this with (4.8) of Lemma 4.2.3 we get

$$\sum_{j \in J_2} C_j \leq \sum_{j \in J_2} (r_j + \tau + p) + \sum_{j \in J_2} (2\tau + \ell_j^* p - \min\{\ell_0^* p, \tau\})$$

$$+ P_2 - P_2^* + n_2 \min\{\ell_0^* p, \tau\}$$

$$\leq 2 \sum_{j \in J_2} C_j^* + P_2 - P_2^* + n_2 \min\{\ell_0^* p, \tau\}.$$

Applying the observation that $P_1 + P_2 = P_1^* + P_2^* = \frac{n(n+1)}{2} p$ we get

$$\sum_{j \in J} C_j = \sum_{j \in J_1} C_j + \sum_{j \in J_2} C_j$$

$$\leq 2\mathsf{Opt} + (P_1 + P_2) - (P_1^* + P_2^*) + n_2 \min\{\ell_0^* p, \tau\} - n_1^2 p$$

$$= 2\mathsf{Opt} + n_2 \min\{\ell_0^* p, \tau\} - n_1^2 p.$$

For the cases $p = 0$, $\ell_0^* \leq n_1 = 0$, or $n_1 \geq n_2$ the difference $n_2 \min\{\ell_0^* p, \tau\} - n_1^2 p$ is at most 0. In general, we can apply that $2\tau \leq C_j^*$ for each job $j \in J_2$. This yields factor 2.5. □

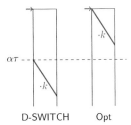

Figure 4.6: Tight example for D-SWITCH with $\Delta = \alpha\tau$: At time $t = 0$ we release k aligned jobs with processing time $p = 0$. Knowing that no further job is released it is optimal to schedule all jobs at time $t = 0$ with objective value of $k \cdot \tau$. Algorithm D-SWITCH waits until time $t = \alpha\tau$ before starting all jobs. This yields a total completion time of $k \cdot (1 + \alpha)\tau$. Hence, we get a ratio of $(1 + \alpha)$ yielding that Lemmas 4.2.4 and 4.2.5 are tight.

Unfortunately, by time $\Delta = \tau$ it is still unknown if $n_1 \geq n_2$ and we have to use the variant according to Lemma 4.2.4. Only if $p = 0$ or $n_1 = 0$ by time $\Delta = \tau$ we can apply Lemma 4.2.5. Figure 4.6 illustrates the tightness of factor $(1 + \sqrt{2})$ if Δ is chosen to be $\sqrt{2}\tau$ and factor 2 for $\Delta = \tau$ and $p = 0$. For the case of $n_1 = 0$ we refer to Figure 4.4. If the complete instance is shifted from $t = 0$ to $t = \tau$ the respective ratio of D-SWITCH and Opt converges to $\frac{4\tau}{2\tau + \varepsilon}$ being arbitrarily close to 2.

Theorem 4.2.6. *Consider bidirectional scheduling on a single segment with identical incompatible jobs. Algorithm D-SWITCH with $\Delta = \sqrt{2}\tau$ has a competitive ratio of $(1+\sqrt{2})$. If $p = 0$ or $n_1 = 0$ it holds that Algorithm D-SWITCH with $\Delta = \tau$ has a competitive ratio of 2.*

5

Competitive-Ratio Approximation Schemes

In this chapter, we present the novel concept of competitive-ratio approximation schemes. To that end, we introduce a new way of designing online algorithms for the example of preemptive parallel machine scheduling. Apart from structuring and simplifying input instances, we find an abstract description of online scheduling algorithms, which allows us to reduce the infinite-size set of all online algorithms to a relevant set of finite size. This is the key for eventually allowing an enumeration scheme that finds an online algorithm with a competitive ratio arbitrarily close to the optimal one. Our method also provides an algorithm to compute the competitive ratio of the designed algorithm, and even the best possible competitive ratio, up to any desired accuracy. We then generalize the ideas to the non-preemptive setting as preparation to ask for a competitive-ratio approximation scheme for bidirectional scheduling.

This chapter is based on joint work with Olaf Maurer, Nicole Megow, and Andreas Wiese [Gün+13].

Competitive analysis is the most popular method for studying the performance of online algorithms. It seeks for online algorithms with preferably best possible competitive ratio. The results of Chapter 4 on bidirectional online scheduling constitute a typical example. Lower bounds on competitive ratios that any online algorithm can achieve are complemented by upper bounds on the competitive ratio of particular algorithms. As long as these values do not coincide the question of a provably optimal online algorithm, w.r.t. competitive analysis, among *all* online algorithms remains open.

One of the few examples admitting optimal online algorithms is total (weighted) completion time minimization for non-preemptive single machine scheduling with 2-competitive online algorithms and a lower bound of 2 [HV96; Phi+98; AP04]. The unweighted preemptive version even achieves an offline optimum in the online setting [Sch68]. For the weighted preemptive setting and multiple parallel machines a long sequence of papers emerged introducing new techniques and algorithms in quest of improved upper and lower bounds. Still, unsatisfactory gaps remain. See Table 5.1 for an overview of the current state-of-the-art in deterministic machine scheduling.

With *competitive-ratio approximation schemes*, we present a novel concept to develop nearly optimal online scheduling algorithms. These approximation schemes compute algorithms with a competitive ratio that is at most a factor $1 + \varepsilon$ larger than the optimal

problem	lower bounds		upper bounds	
$1\|r_j, pmtn\|\sum C_j$	1		1	[Sch68]
$1\|r_j, pmtn\|\sum w_j C_j$	1.073	[ES03]	1.566	[Sit10a]
$1\|r_j\|\sum C_j$	2	[HV96]	2	[HV96]
$1\|r_j\|\sum w_j C_j$	2	[HV96]	2	[AP04]
$P\|r_j, pmtn\|\sum C_j$	1.047	[Ves97]	1.25	[Sit10b]
$P\|r_j, pmtn\|\sum w_j C_j$	1.047	[Ves97]	1.791	[Sit10b]
$P\|r_j\|\sum w_j C_j$	1.309^1	[Ves97]	1.791	[Sit10b]
$R\|r_j\|\sum w_j C_j$	1.309	[Ves97]	8	[Hal+97]

Table 5.1: Lower and upper bounds on the competitive ratio of deterministic online algorithms for (parallel) machine scheduling.

ratio ρ^* for any $\varepsilon > 0$. Moreover, our method also provides an algorithm to compute the competitive ratio of the designed algorithm, and even the best possible competitive ratio, up to any desired accuracy. Hence, we define a competitive-ratio approximation scheme as a procedure that computes a nearly optimal online algorithm and at the same time provides a nearly exact estimate of the optimal competitive ratio:

Definition 5.0.1. A *competitive-ratio approximation scheme* computes for a given $\varepsilon > 0$ an online algorithm A with a competitive ratio $\rho_A \leq (1 + \varepsilon)\rho^*$. Moreover, it determines a value ρ' such that $\rho' \leq \rho^* \leq (1 + \varepsilon)\rho'$.

We present our technique focusing on preemptive parallel machine scheduling for any constant number of machines m, i.e., $Pm|r_j, pmtn|\sum w_j C_j$. Afterward, we extend the presented technique to the corresponding non-preemptive setting as preparation to ask for a competitive-ratio approximation scheme for bidirectional scheduling.

The results can additionally be extended to related machines (assuming a constant range of machine speeds in the non-preemptive case), preemptive scheduling on unrelated machines, and other cost functions such as the makespan, $C_{\max}$ and $\sum_{j \in J} w_j f(C_j)$ where f is an arbitrary monomial function with fixed exponent, cf. [Gün+13]. Also randomized competitive-ratio approximation schemes are obtained.

After the first publication of the new concept further subsequent work extending our methods was published. Kurpisz et al. [Kur+13] present competitive-ratio approximation schemes for minimizing the makespan on non-preemptive scheduling on unrelated machines, for job shop scheduling, and for scheduling on a single machine with delivery times. Chen et al. [Che+15] present such a scheme for the problem of scheduling jobs one by one on unrelated machines, the objective function being to minimize the makespan or the L_p-norm of the machine loads. For the case of identical machines and the makespan objective they even bound the running time by a polynomial in the number of machines. Independently, for the same online model Megow and Wiese found a competitive-ratio approximation scheme for minimizing the makespan on identical and related machines [MW13]. They do not provide a polynomial bound on the running time but their construction is more compact than the one in [Che15]. Finally, Mömke pro-

---

[1]For $m = 1, 2, 3, 4, 5, \ldots 100$ the lower bound is $LB = 2, 1.520, 1.414, 1.373, 1.364, \ldots 1.312$.

vides a competitive-ratio approximation scheme for the k-server problem in fixed finite metrics [Möm13].

To achieve our results, we introduce a new and unusual way of designing online scheduling algorithms. We present an abstraction in which online algorithms are formalized as *algorithm maps* (Section 5.2). Such a map receives as input a set of unfinished jobs together with the schedule computed so far. Based on this information, it returns a schedule for the next time instant. This view captures exactly how online algorithms operate under limited information. The total number of algorithm maps is unbounded. However, we show that there is a finite subset which approximates the entire set. More precisely, for any algorithm map there is a map in our subset whose competitive ratio is at most by a factor $1 + \varepsilon$ larger. To achieve this reduction, we first apply several standard techniques (see e.g. [Afr+99]), such as geometric rounding, time-stretch, and weight-shift, to transform and simplify the input problem without increasing the objective value too much (Section 5.1.1). The key, however, is the insight that it suffices for an online algorithm to base its decisions on the currently unfinished jobs and a very *limited part* of the so far computed schedule—rather than the entire history (Section 5.1.2). This allows for an enumeration of all relevant instances and algorithm maps.

To summarize, our algorithmic scheme outputs (1) an online algorithm with competitive ratio arbitrarily close to the best possible one and (2) up to a $1 + \varepsilon$ factor the corresponding value. This implies a $(1 + \varepsilon)$-estimate for the optimal competitive ratio. The approach contrasts strongly those results where (matching) upper and lower bounds on the competitive ratio of a particular and of all online algorithms were derived *manually*, instead of executing an algorithm using, e.g., a computer. In general, there are no computational means to enumerate all possible online algorithms and instances or even to determine the competitive ratio of a single algorithm. We overcome these issues and reduce the long-time ongoing search for the best possible competitive ratio for the considered problems to a question that can be answered by a finite algorithm.

Although the enumeration scheme for identifying the (nearly) optimal online algorithm heavily exploits unbounded (but finite) computational resources, the resulting algorithm itself has polynomial running time. As a consequence, there are efficient online algorithms for the considered problems with almost optimal competitive ratios. Hence, the granted additional, even unbounded, computational power of online algorithms does not yield any significant benefit here.

5.1 General simplifications and techniques

We start by discussing several transformations that simplify the input and reduce the structural complexity of online schedules for $\text{Pm}\,|\,r_j, pmtn\,|\sum w_j C_j$. Our construction combines several transformation techniques known for offline PTASs (see [Afr+99] and the references therein) and a new technique to subdivide an instance online into parts which can be handled separately.

We will use the terminology that *at $1 + \mathcal{O}(\varepsilon)$ loss we can restrict* to instances or schedules with certain properties. This means that we lose at most a factor $1 + \mathcal{O}(\varepsilon)$, as $\varepsilon \to 0$, by limiting our attention to those. We bound several relevant parameters by constants. If not stated differently, any mentioned constant depends only on ε and m.

To begin with, we use the standard simplification technique of *geometric rounding* introduced in [Afr+99]. For the sake of completeness, we prove the required properties in the following lemma.

Lemma 5.1.1. *At $1 + O(\varepsilon)$ loss we can restrict to instances where all processing times, release dates, and weights are powers of $1 + \varepsilon$, no job is released before time $t = 1$, and $r_j \geq \varepsilon \cdot p_j$ for all jobs j.*

Proof. We prove that any given schedule can be adapted at $1 + O(\varepsilon)$ loss such that the required properties can be assumed for the corresponding instance.

First, we increase each weight w_j to the smallest power of $(1 + \varepsilon)$ not smaller than w_j which yields a value with $(1 + \varepsilon)^x = (1 + \varepsilon)(1 + \varepsilon)^{x-1} < (1 + \varepsilon)w_j$.

Within a given preemptive schedule, the job volume of each job j is assigned to a set of time intervals on machines (such that their lengths sum up to p_j). Multiplying the boundary values of each interval by $(1 + \varepsilon)$ results in time intervals assigned to each j with a total length of $(1 + \varepsilon)p_j$. Hence, we get a feasible schedule even when rounding up all processing times

Consider now a schedule with rounded processing times. We again multiply each boundary value of the intervals by $(1 + \varepsilon)$ and shift the processing volume of each job j to the latest possible time intervals within its assigned intervals. Then, job j completes at time $(1 + \varepsilon)C_j$, and the earliest point in time at which j is processed increases from S_j to $S_j + \varepsilon p_j$. Hence, by losing at most a factor $(1 + \varepsilon)$ we may assume that each job j has a release date $r_j \geq \varepsilon p_j$. If necessary, the parameters of all jobs can be scaled by some power of $(1 + \varepsilon)$ such that the earliest release date is at least one (since jobs with $r_j = p_j = 0$ can be ignored).

With a similar reasoning, we can finally round at a loss of $(1 + \varepsilon)$ each release date to the next power of $(1 + \varepsilon)$. □

The geometric rounding procedure allows us to see intervals of the form $I_x := [R_x, R_{x+1})$ with $R_x := (1 + \varepsilon)^x$ as atomic entities. Note that $|I_x| = \varepsilon \cdot R_x$. An online algorithm can define the corresponding schedule at the beginning of an interval since no further jobs are released until the next interval. Moreover, we make, at $1 + \varepsilon$ loss, the simplifying assumption that each job j finishing within I_x contributes $w_j \cdot D_j$ to the objective function where $D_j := R_{x+1}$.

5.1.1 Simplification within intervals

The goal of this section is to reduce the number of situations that can arise at the beginning of an interval. For each ε we identify constantly many relevant inputs per interval. And for a given input, we reduce the number of relevant algorithmic actions within each interval to a constant.

To do so, we use the techniques of *time-stretching* and partitioning jobs released at time R_x into *large and small* jobs, see [Afr+99]. In an online interpretation of time-stretching, we shift the work assigned to any interval I_x to the interval I_{x+1}. When doing this operation once we speak of *one time-stretch*. This can be done at a loss of $1 + \varepsilon$ and we obtain free space of size $\varepsilon \cdot |I_x|$ in each interval I_{x+1}. For each $x \geq 0$ we define $L_x = \{j \in J \mid r_j = R_x, p_j > \varepsilon^2 |I_x|\}$ to be the set of *large* jobs released at R_x, and we let $S_x = \{j \in J \mid r_j = R_x, p_j \leq \varepsilon^2 |I_x|\}$ be the set of *small* jobs released at R_x.

We first take care of the small jobs. Since they are small, there is a lot of flexibility in scheduling them. We show that we can, at a small loss, fix in advance the order in which the jobs in each S_x are processed. This enables us to group very small jobs to job *packs*. Treating each pack as a single job we then get a lower bound on the processing time of the released jobs.

Recall that the Smith ratio of a job j is defined as w_j/p_j. A non-increasing ordering by the Smith ratios is an ordering according to *Smith's rule* [Smi56]. We say that a job is *partially processed* at some point in time if it has been processed, but not yet completed. For a set of jobs J we define $p(J) := \sum_{j \in J} p_j$ and $w(J) := \sum_{j \in J} w_j$.

Lemma 5.1.2. *At $1 + \varepsilon$ loss we can restrict to schedules such that for each interval I_x the small jobs scheduled within this interval are chosen by Smith's rule from the set $S_{\leq x} := \bigcup_{x' \leq x} S_{x'}$ and no small job is preempted or only partially processed at the end of an interval. Therefore, we can restrict to instances with $p(S_x) \leq m \cdot |I_x|$ for each interval I_x.*

Proof. Consider some schedule S and apply one time-stretch. The resulting schedule S' has a free space of $\varepsilon |I_x|$ in each interval I_{x+1}, and its total cost is within a factor $(1 + \varepsilon)$ of the cost of S. Remove now all small jobs from each interval I_{x+1}. For each x and each job j we denote by $q_{x,j}$ the amount of processing time of j assigned by S' to interval I_x. Furthermore, let Q_x be the sum of $q_{x,j}$ over all small jobs and Q'_x the respective sum over all large jobs. By A_x we denote the set of all removed small jobs completed by S' in I_x. Now, apply for each $x \geq 0$ the following procedure consecutively: First, we remove all large jobs from interval I_x and reschedule them machine-wise on the first $|I_{x-1}|$ units of processing time, that is, we fill the machines consecutively and with preemption until an amount of $q_{x,j}$ is assigned for each large job j which fits due to $Q'_x \leq m \cdot |I_{x-1}|$. Furthermore, the costs are not increased by this procedure. We now have a number of $k \geq 0$ machines in I_x whose first $|I_{x-1}|$ units of processing time are completely filled and $m - k$ machines where this is not the case, confer the respective strategy of McNaughton [McN59].

Now, fill the idle time of each machine $i = k + 1, \ldots, m$ with unscheduled jobs from $S_{\leq x}$ chosen in the order of their Smith ratio without preemption until an amount of Q_x is achieved or no further small job is available, where a switch to the next machine is accomplished as soon as $|I_{x-1}|$ units of processing are covered on each machine. Again, this is possible since $Q_x + Q'_x \leq m \cdot |I_{x-1}|$. The procedure increases the processing time on the latter machine by at most $\varepsilon^2 |I_x|$, the maximum size of such jobs, which is smaller than the created extra space of $\varepsilon |I_{x-1}|$ for sufficiently small ε. Denote by B_x the set of small jobs assigned to I_x in this step.

Inductively, we can prove that for each I_x the total weight of jobs from $S_{\leq x}$ that are completed before the end of I_x has not decreased compared to S'. For the first interval I_{x_1} with $Q_{x_1} > 0$ it holds that $w(A_{x_1}) \leq w(B_{x_1})$ since the jobs of B_{x_1} are assigned according to Smith's rule and $p(B_{x_1}) \geq Q_{x_1} \geq p(A_{x_1})$. This proves the base case since each job of B_{x_1} is completed before the end of I_{x_1}. For the induction step we use that $w(A_{\leq x}) \leq w(B_{\leq x})$ with $A_{\leq x} := \bigcup_{x' \leq x} A_{x'}$ and $B_{\leq x}$ respectively for some x. Assume by contradiction $w(A_{\leq \tilde{x}}) > w(B_{\leq \tilde{x}})$ for the next $\tilde{x} > x$ with $Q_{\tilde{x}} > 0$. This implies that

$$w(A_{\tilde{x}}) > w(B_{\leq \tilde{x}}) - w(A_{\leq x}) + w(B_{\tilde{x}}).$$

Intuitively, the total weight of jobs in $A_{\tilde{x}}$ is higher than the surplus weight gained in the earlier intervals (being non-negative) together with the total weight of $B_{\tilde{x}}$. This contradicts the fact that we assigned for each $x' \leq \tilde{x}$ jobs with a total processing time of $p(B_{\leq x'}) \geq p(A_{\leq x'})$ via Smith's rule. Hence, we get $w(A_{\leq \tilde{x}}) \leq w(B_{\leq \tilde{x}})$. The induction hypothesis follows since each job in $B_{\tilde{x}}$ is completed before the end of $I_{\tilde{x}}$. Recall that we modified the objective function by rounding up the completion times of jobs to the end of the respective interval. Consequently, the observation on the completed weights before each interval end yields that the objective value is not increased by the described reassignment (cf. Lemma 4.1.4, p. 70). Furthermore, the procedure ensures that no small job is preempted and that any small job finishes in the same interval where it started.

Since we can now assume, that the jobs of each S_x are chosen to be scheduled within I_x in non-increasing order of their Smith ratios (completely and without preemption), we can conclude the last claim of the lemma. Note that the total processing time in interval I_x is $m|I_x|$. Pick the jobs of S_x in the respective order until the total processing time of picked jobs just exceeds $m|I_x|$. The remaining jobs of S_x cannot be scheduled within I_x and hence, by the above argument we can safely move their release dates to R_{x+1}. $\qquad\square$

For the grouping of very small jobs to a pack we need the following small observation stating that the extra space of one interval created by one time-stretch on a machine is sufficient to cover one small job of each earlier Interval.

Lemma 5.1.3. $\sum_{x<y} \varepsilon^2 |I_x| \leq \varepsilon |I_y|$

Proof. To prove the claim we use that the n-th partial sum $\sum_{k=0}^{n} az^k$ of a geometric series equals $a\frac{1-z^{n+1}}{1-z}$:

$$\varepsilon^3 \sum_{x<y} (1+\varepsilon)^x = \varepsilon^3 \frac{1-(1+\varepsilon)^y}{1-(1+\varepsilon)} = \varepsilon^2((1+\varepsilon)^y - 1) \leq \varepsilon |I_y|$$

$\qquad\square$

Lemma 5.1.4. *At $1 + \mathcal{O}(\varepsilon)$ loss we can restrict to instances such that $p_j \geq \frac{\varepsilon^2}{4} \cdot |I_x|$ for each job $j \in S_x$. In these instances, the number of distinct processing times of each set S_x is bounded from above by $\log_{1+\varepsilon} 4$.*

Proof. We call a job $j \in S_x$ *tiny* if $p_j \leq \frac{\varepsilon^2}{4} \cdot |I_x|$. Let $T_x = \{j_1, j_2, ..., j_{|T_x|}\}$ denote all tiny jobs released at R_x. W.l.o.g. assume that they are ordered non-increasingly by their Smith ratios w_j/p_j. Let ℓ be the largest integer such that $\sum_{i=1}^{\ell} p_i \leq \frac{\varepsilon^2}{2} \cdot |I_x|$. We define the pack $P_x^1 := \{j_1, ..., j_\ell\}$. We define the processing time of pack P_x^1 to be $\sum_{i=1}^{\ell} p_i$ and its weight to be $\sum_{i=1}^{\ell} w_i$. We continue iteratively until we assigned all tiny jobs to packs. By definition of the processing time of tiny jobs, the processing time of all but possibly the last pack released at time R_x is in the interval $\left[\frac{\varepsilon^2}{4} \cdot |I_x|, \frac{\varepsilon^2}{2} \cdot |I_x|\right]$.

We apply one time-stretch. In any schedule that assigns small jobs according to Smith's rule to intervals there is for each I_x at most one partially processed job pack per machine from each of the previous release dates $R_{x'} < R_x$. Since $\sum_{x'<x} \varepsilon^2 |I_{x'}| \leq \varepsilon |I_x|$ due to Lemma 5.1.3, we can schedule all of them in the newly created space. This also includes space to increase the processing time of the very last pack of each $S_{x'}$ to $\frac{\varepsilon^2}{4} \cdot |I_{x'}|$, if necessary. Therefore, we can enforce that at $1 + O(\varepsilon)$ loss all tiny jobs of the same pack are scheduled in the same interval on the same machine. Hence, we can treat each pack as a single job whose processing time and weight matches the respective values of the pack.

Finally, at $1 + O(\varepsilon)$ loss we can ensure that the processing times and weights of the new jobs (which still remain small) are powers of $1 + \varepsilon$. Consequently, the processing times of jobs in S_x are of the form $(1 + \varepsilon)^y$ within the following range:

$$\frac{e^3}{4} \cdot (1 + \varepsilon)^x \le (1 + \varepsilon)^y \le \varepsilon^2 |I_x| = \varepsilon^3 (1 + \varepsilon)^x.$$

The number of integers y satisfying these inequalities is bounded from above by $\log_{1+\varepsilon} 4$.
□

For the large jobs released at the beginning of an interval we obtain an upper bound on their length by their relation of release date and processing time (Lemma 5.1.1). This induces a constant upper bound on the number of occurring processing times.

Lemma 5.1.5. *The number of distinct processing times of jobs in each set L_x is bounded from above by $4 \log_{1+\varepsilon} \frac{1}{\varepsilon}$.*

Proof. Let $j \in L_x$ be a large job released at R_x with processing time $p_j = (1+\varepsilon)^y > \varepsilon^2 |I_x|$ for some integer y. By Lemma 5.1.1, we know that $p_j \le \frac{1}{\varepsilon} r_j = \frac{1}{\varepsilon} (1 + \varepsilon)^x$ and hence,

$$\varepsilon^3 (1 + \varepsilon)^x = \varepsilon^2 |I_x| < (1 + \varepsilon)^y \le \frac{1}{\varepsilon} (1 + \varepsilon)^x.$$

The number of integers y which satisfy the above inequalities is upper-bounded by the constant claimed in the lemma.
□

We say that two large jobs are of the same *type* if they have the same processing time and the same release date. By an exchange argument, we can restrict ourselves without any loss to schedules in which at each point in time at most m large jobs of each type are partially scheduled. Since the amount of work that can be processed within each interval is bounded, the number of large jobs of the same type can also be bounded.

Lemma 5.1.6. *Without loss, we can restrict to instances such that for each set L_x:*

$$|L_x| \le (m/\varepsilon^2 + m) 4 \log_{1+\varepsilon} \frac{1}{\varepsilon}$$

Proof. Let $L_{x,p} \subseteq L_x$ denote the set of jobs in L_x with processing time p, i.e., they are of the same type. Since $p_j > \varepsilon^2 |I_x|$ for each job $j \in L_x$, at most $m/\varepsilon^2 + m$ jobs in $L_{x,p}$ can be started before I_{x+1}. By an exchange argument we can assume that they are among the $m/\varepsilon^2 + m$ jobs with the largest weight in $L_{x,p}$. Hence, the release date of all other jobs in $L_{x,p}$ can be moved to R_{x+1} without any cost. By Lemma 5.1.5 there are at most $4 \log_{1+\varepsilon} \frac{1}{\varepsilon}$ distinct processing times p of large jobs in L_x and, thus, the claim follows.
□

As we simplified the objective function by pretending all jobs to complete at the end of an interval, the only information needed for computing the objective function value is the interval in which a job completes. The only part that has not been bounded yet is the amount that large jobs are processed within the single intervals.

Lemma 5.1.7. *There is a constant $\mu \in \mathbb{N}$ such that at $1 + O(\varepsilon)$ loss we can restrict to schedules such that at the end of each interval, each large job j is processed to an extent which is an integer multiple of p_j/μ.*

Proof. We apply one time-stretch and choose $\mu \in \mathbb{N}$ to be a constant such that $1/\mu$ is smaller than $\varepsilon^4/(8 \log_{1+\varepsilon} \frac{1}{\varepsilon})$. Consider now an interval I_x and the jobs that are scheduled in I_x before the time-stretch. For each job $j \in L_{x'}, x' \leq x$ that was partially processed at time R_{x+1} we must extend the amount of time j is processed within I_{x+1} by at most p_j/μ to achieve the stated property. Using Lemma 5.1.1 this value can be bounded by

$$\frac{p_j}{\mu} \leq \frac{R_{x'}}{\varepsilon} \cdot \frac{\varepsilon^4}{8 \log_{1+\varepsilon} \frac{1}{\varepsilon}} = \frac{\varepsilon^2 |I_{x'}|}{2 \cdot 4 \log_{1+\varepsilon} \frac{1}{\varepsilon}}.$$

Recall that we can assume without any loss that at the end of each interval at most one large job per job type is partially processed on each machine. Since the number of job types of $L_{x'}$ is bounded by $4 \log_{1+\varepsilon} \frac{1}{\varepsilon}$ (Lemma 5.1.5), the space of $\varepsilon^2 |I_{x'}|/2$ is sufficient to handle each job type of $L_{x'}$. Applying $\sum_{x' < x} \varepsilon^2 |I_{x'}| \leq \varepsilon |I_x|$, we see that this amount of free space was created by the time-stretch. □

To summarize, we get the following simplifications within intervals concerning input and scheduling decisions. Recall that we can calculate the value of μ that depends only on ε. Furthermore, we define Λ to be a constant upper bound on the number of released jobs in each interval, i.e., Λ is an integer of at most $\lceil \frac{4m}{\varepsilon^2} + 4 \log_{1+\varepsilon} \frac{1}{\varepsilon} (\frac{m}{\varepsilon^2} + m) \rceil$.

Corollary 5.1.8. *At $1 + \mathcal{O}(\varepsilon)$ loss we can assume that for each interval I_x*
(a) each job j released at time R_x has a processing time $p_j = (1 + \varepsilon)^k \in [\frac{\varepsilon^3}{4} R_x, \frac{1}{\varepsilon} R_x]$ for some integer k,
(b) there are at most $\log_{1+\varepsilon}(4/\varepsilon^4)$ distinct processing time values of jobs released at R_x,
(c) at most Λ jobs are released at R_x,
(d) each small job starting in I_x completes in I_x without preemption, and
(e) at the end of I_x, each job j is processed to an extent of $\ell_{x,j} \cdot p_j/\mu$ for some $\ell_{x,j} \in \{0, \ldots, \mu\}$.

5.1.2 Irrelevant history

The schedule for an interval returned by an online algorithm may depend on the set of currently unfinished jobs and possibly the entire schedule computed so far. In the remainder of this section we show why we can assume that an online algorithm only takes a finite amount of history into account in its decision making, namely, the jobs with relatively large weight released in the last constantly many intervals.

Firstly, we show that we may assume that each job completes within constantly many intervals after its release.

Lemma 5.1.9. *There is a constant s such that at $1 + \mathcal{O}(\varepsilon)$ loss we can restrict to schedules such that for each interval I_x there is a subinterval of I_{x+s-1} which is large enough to process all jobs released at R_x and during which only those jobs are executed. We call this subinterval the* safety net *of interval I_x. We can assume that each job released at R_x finishes before time R_{x+s}.*

Proof. By using Lemmas 5.1.1, 5.1.2, and 5.1.6 we get

$$p(S_x) + p(L_x) \leq m \cdot |I_x| + (m/\varepsilon^2 + m) \cdot \left(4\log_{1+\varepsilon} \frac{1}{\varepsilon} \right) \cdot \frac{1}{\varepsilon} (1 + \varepsilon)^x$$

$$\leq m \cdot (1 + \varepsilon)^x \left(\varepsilon + \frac{8}{\varepsilon^3} \log_{1+\varepsilon} \frac{1}{\varepsilon} \right)$$

$$= \varepsilon \cdot |I_{x+s-2}|$$

for a suitable constant s, depending on ε and m. Stretching time once, we gain enough free space at the end of each interval I_{x+s-1} to establish the safety net for each job set $p(S_x) + p(L_x)$. $\qquad\square$

Given the bound on the number of intervals between release and completion times of jobs, we partition the time horizon into periods such that no job is "alive" for more than two periods. For each integer $k \geq 0$, we define *period* Q_k to consist of the s consecutive intervals $I_{k \cdot s}, ..., I_{(k+1) \cdot s - 1}$. We add an artificial period Q_{-1} for the interval $[0, 1)$ in which no job is released. Hence, we can assume by Lemma 5.1.9 that each job released in period Q_k is completed by the end of period Q_{k+1}. For ease of notation, we will treat a period Q as the set of jobs released in that period. For a set of jobs J we denote by $rw(J) := \sum_{j \in J} r_j w_j$ their *release weight*. Note that $rw(J)$ forms a lower bound on the quantity that these jobs must contribute to the objective value in *any* schedule. Due to Lemma 5.1.9, we also obtain an upper bound of $(1 + \varepsilon)^s \cdot rw(J)$ for the latter quantity.

We will now determine at the end of each period how important its released jobs are compared to the previous periods. If the weights released in a series of periods grow large enough from period to period we will see that the overall objective value is dominated by the contribution of the constantly many last periods. If otherwise the weights of a period are too low compared to the preceding ones and its jobs can be moved to their safety net with a small loss, we will see that the following periods can be treated independently. To this end, we define that $p > 0$ consecutive periods $Q_k, ..., Q_{k+p-1}$ are a *sequence of significant periods* if $rw(Q_{k+\ell}) > \frac{\varepsilon}{(1+\varepsilon)^s} \cdot \sum_{i=0}^{\ell-1} rw(Q_{k+i})$ for each $\ell = 1, ..., p - 1$. This implies exponential growth for the series of partial sums of release weights and we can prove the dominance of a few of the last periods.

Lemma 5.1.10. *There is a constant K such that for each sequence $Q_k, Q_{k+1}, ..., Q_{k+p-1}$ of significant periods we have that*

$$\sum_{i=0}^{p-K-1} (1 + \varepsilon)^s \, rw(Q_{k+i}) \leq \varepsilon \cdot \sum_{i=p-K}^{p-1} rw(Q_{k+i}).$$

Proof. Let $\delta := \frac{\varepsilon}{(1+\varepsilon)^s}$. Since we consider a sequence of significant periods, we get

$$rw(Q_{k+\ell}) > \delta \cdot \sum_{i=0}^{\ell-1} rw(Q_{k+i}) \qquad \forall \ell = 1, ..., p - 1.$$

This implies that

$$(1 + \delta)rw(Q_{k+\ell}) > \delta \cdot \sum_{i=0}^{\ell} rw(Q_{k+i}) \qquad \forall \ell = 1, ..., p - 1.$$

Hence, we get

$$\frac{\sum_{i=0}^{\ell-1} rw(Q_{k+i})}{\sum_{i=0}^{\ell} rw(Q_{k+i})} = 1 - \frac{rw(Q_{k+\ell})}{\sum_{i=0}^{\ell} rw(Q_{k+i})} < 1 - \frac{\delta}{1+\delta} = \frac{1}{1+\delta} < 1 \quad \forall \ell = 1, \dots, p-1$$

which implies

$$(5.1) \qquad \sum_{i=0}^{\ell-1} rw(Q_{k+i}) < \frac{1}{1+\delta} \sum_{i=0}^{\ell} rw(Q_{k+i}) \qquad \forall \ell = 1, \dots, p-1.$$

In other words, if we remove $Q_{k+\ell}$ from $\cup_{i=0}^{\ell} Q_{k+i}$, the total release weight of the set decreases by a factor of at least $1/(1+\delta) < 1$. Recursively applying (5.1) we get for any K

$$\sum_{i=0}^{p-1-K} rw(Q_{k+i}) < \left(\frac{1}{1+\delta}\right)^K \sum_{i=0}^{p-1} rw(Q_{k+i})$$

$$= \frac{1}{(1+\delta)^K} \left(\sum_{i=0}^{p-K-1} rw(Q_{k+i}) + \sum_{i=p-K}^{p-1} rw(Q_{k+i}) \right)$$

and hence

$$\left(1 - \left(\frac{1}{(1+\delta)^K}\right)\right) \sum_{i=0}^{p-K-1} rw(Q_{k+i}) < \frac{1}{(1+\delta)^K} \sum_{i=p-K}^{p-1} rw(Q_{k+i}).$$

To ensure that $(1+\varepsilon)^s \sum_{i=0}^{p-K-1} rw(Q_{k+i}) < \varepsilon \cdot \sum_{i=p-K}^{p-1} rw(Q_{k+i})$ we choose K sufficiently large such that

$$\frac{1}{1 - \frac{1}{(1+\delta)^K}} \cdot \frac{1}{(1+\delta)^K} = \frac{1}{(1+\delta)^K - 1} < \delta = \frac{\varepsilon}{(1+\varepsilon)^s}.$$

Hence, K is some constant integer larger than $\log_{1+\delta}(1/\delta + 1)$ and depends only on ε. $\qquad\square$

Using the safety net (Lemma 5.1.9) the result implies that an ε-fraction of the weighted completion time of the last $K - 1$ periods of a sequence of significant periods is an upper bound on the weighted completion time of the previous periods of this sequence. Therefore, the objective value is dominated by the contribution of the last $K - 1$ periods. We will need this later to show that at $1 + \mathcal{O}(\varepsilon)$ loss we can assume that an online algorithm bases its decisions only on a constant amount of information.

To do so, we consider in the following a sequence of significant periods $Q_k, \dots, Q_{k+p-1}$ and an interval I_x of period Q_{k+p} where the new released jobs are just revealed to an online algorithm. Recall that the online algorithm does not know the release weight of the current period Q_{k+p} unless I_x is the last interval of Q_{k+p}. Nevertheless, we know by Lemma 5.1.10 that the costs of $Q_k, \dots, Q_{k+p-1}$ are dominated by the last $K - 1$ periods. Hence, the costs until interval I_x are dominated by the last important $\Gamma := Ks$ intervals including I_x. In addition to the jobs that have been released very early, also the jobs with very small weight in comparison to at least one other job can be almost neglected for the total costs. Hence, we partition the jobs into relevant and irrelevant jobs using these two criteria.

Definition 5.1.11. Let J be a set of jobs. A job $j \in J$ with $r_j \leq R_x$ is called *recent at time R_x* if $R_{x-\Gamma} \leq r_j$. Otherwise, it is called *old at time R_x*. Job j is furthermore called *dominated at time R_x* if it is dominated at time R_{x-1} or if there is a job $j' \in J$ being recent and not dominated at time R_x such that $w_j < \frac{\varepsilon}{\Lambda \cdot \Gamma \cdot (1+\varepsilon)^{\Gamma+s}} w_{j'}$. Now, job j is *irrelevant at time R_x* if it is old or dominated at time R_x and otherwise *relevant at time R_x*. Denote the respective subsets of a job set J by $\mathrm{Rec}_x(J)$, $\mathrm{Old}_x(J)$, $\mathrm{Dom}_x(J)$, $\mathrm{Ir}_x(J)$, and $\mathrm{Rel}_x(J)$.

The subsequent lemma states that the irrelevant jobs can almost be ignored for the objective value of a schedule even when scheduled in their safety nets. This implies that we can restrict at $1 + O(\varepsilon)$ loss to online algorithms which schedule the remaining part of a job in its safety net, once it has become irrelevant.

Lemma 5.1.12. *Let $Q_k, \ldots, Q_{k+p-1}$ be a sequence of significant periods and let J be the jobs of $Q_k, \ldots, Q_{k+p}$ released by the beginning of an interval $I_x \in Q_{k+p}$. Then*

$$(5.2) \qquad (1+\varepsilon)^s rw(\mathrm{Ir}_x(J)) \leq 6\varepsilon \cdot rw(\mathrm{Rel}_x(J)).$$

Proof. By definition, an irrelevant job at time R_x is either old or dominated. Hence, $\mathrm{Ir}_x(J)$ is the disjoint union of $\mathrm{Old}_x(J)$ and $\mathrm{Dom}_x(J) \cap \mathrm{Rec}_x(J)$.

We start by giving a bound on the recent but dominated jobs and define $rw_{\max} := \max\{rw(j) \mid j \in \mathrm{Rel}_x(J) \cup \mathrm{Old}_x(J)\}$. Consider a job $j \in J$ that is dominated and recent at time R_x. By definition, there is a time $r_j \leq R_{x'} \leq R_x$ at which another job j' that is relevant at time $R_{x'}$ dominates j. Choose $R_{x'}$ to be as late as possible. As j' is recent at time $R_{x'}$, we get by $R_{x'}(1+\varepsilon)^{-\Gamma} \leq r_{j'}$ that $r_j \leq R_{x'} \leq (1+\varepsilon)^\Gamma r_{j'}$. Because we chose $R_{x'}$ to be as late as possible, we can assume that j' is not dominated at time R_x. Hence, j' is either relevant or old at time R_x. Therefore, there is for each $j \in \mathrm{Dom}_x(J) \cap \mathrm{Rec}_x(J)$ a job j' in $\mathrm{Rel}_x(J) \cup \mathrm{Old}_x(J)$ such that

$$w_j r_j \leq \frac{\varepsilon}{\Lambda \cdot \Gamma \cdot (1+\varepsilon)^{\Gamma+s}} w_{j'} r_j$$

$$\leq \frac{\varepsilon}{\Lambda \cdot \Gamma \cdot (1+\varepsilon)^{\Gamma+s}} w_{j'} r_{j'} (1+\varepsilon)^\Gamma$$

$$\leq \frac{\varepsilon}{\Lambda \cdot \Gamma \cdot (1+\varepsilon)^s} rw_{\max}.$$

Since at most Λ jobs are released at the beginning of each interval (Corollary 5.1.8) and $rw_{\max}$ is the release weight of a job in $\mathrm{Rel}_x(J) \cup \mathrm{Old}_x(J)$ we get:

$$(1+\varepsilon)^s rw(\mathrm{Dom}_x(J) \cap \mathrm{Rec}_x(J)) = (1+\varepsilon)^s \sum_{j \in \mathrm{Dom}_x(J) \cap \mathrm{Rec}_x(J)} w_j r_j$$

$$\leq (1+\varepsilon)^s \Lambda \cdot \Gamma \frac{\varepsilon}{\Lambda \cdot \Gamma \cdot (1+\varepsilon)^s} rw_{\max}$$

$$= \varepsilon \cdot rw_{\max}$$

$$\leq \varepsilon \left(rw(\mathrm{Rel}_x(J)) + rw(\mathrm{Old}_x(J)) \right).$$

Moreover, Lemma 5.1.10 implies for the old jobs at time R_x that

$$(1+\varepsilon)^s rw(\mathrm{Old}_x(J)) \leq \varepsilon \cdot rw(\mathrm{Rec}_x(J))$$

$$= \varepsilon \cdot \left(rw(\mathrm{Rel}_x(J)) + rw(\mathrm{Dom}_x(J) \cap \mathrm{Rec}_x(J)) \right).$$

Combining all these arguments yields

$$(1 + \varepsilon)^s rw(\mathrm{Ir}_x(J)) = (1 + \varepsilon)^s \left[rw(\mathrm{Old}_x(J)) + rw(\mathrm{Dom}_x(J) \cap \mathrm{Rec}_x(J))\right]$$
$$\leq \varepsilon \cdot rw(\mathrm{Rel}_x(J)) + 2\varepsilon \cdot \left[rw(\mathrm{Rel}_x(J)) + rw(\mathrm{Old}_x(J))\right]$$
$$\leq 3\varepsilon \cdot rw(\mathrm{Rel}_x(J)) + 2\varepsilon \cdot rw(\mathrm{Ir}_x(J)).$$

With $\varepsilon < \frac{1}{3}$ we obtain the bound $(1 + \varepsilon)^s rw(\mathrm{Ir}_x(J)) \leq 6\varepsilon \cdot rw(\mathrm{Rel}_x(J))$. □

With the preceding lemmas we have identified for each point in time a subset of jobs that is relevant at this time—under the assumption that we consider a sequence of significant periods $Q_k, \ldots, Q_{k+p-1}$. It now remains to handle the case that the following period Q_{k+p} is not significant. In this case, we know that $(1 + \varepsilon)^s rw(Q_{k+p}) \leq \varepsilon \cdot \sum_{i=0}^{p-1} rw(Q_{k+i})$. Hence, completing all unfinished jobs of Q_{k+p} in their safety nets costs only an ε-fraction of the costs caused by the preceding significant periods. And since an online algorithm is allowed to preempt these jobs it can now schedule the succeeding periods independently. However, in the following we will state a more general result which will be useful to eventually proving that online algorithms can forget all irrelevant jobs.

Consider for some $p > 0$ a set of consecutive periods $Q_k, \ldots, Q_{k+p}$. We define this set to be a *part* if $Q_k, \ldots, Q_{k+p-1}$ is a sequence of significant periods and if $(1 + \varepsilon)^s rw(Q_{k+p}) \leq 8\varepsilon \cdot \sum_{i=0}^{p-1} rw(Q_{k+i})$. We then call Q_{k+p} to be *insignificant* (even though it is possible that Q_{k+p} is still significant). We now consider a partitioning of a given instance $\mathcal{I}$ into parts $P_i, i = 0, \ldots, \ell$, and denote the insignificant period of each part P_i by $Q_{a_{(i+1)}}$. With $a_0 := -1$ each part P_i consists of the periods $Q_{a_i+1}, \ldots, Q_{a_{(i+1)}}, i = 0, \ldots, \ell$. Again, we identify with P_i all jobs released in this part.

With the possibility to preempt jobs, we now treat each part P_i of a partition as a separate instance that we present to a given online algorithm. For the final output, we concatenate the computed schedules for the different parts. By the following lemma, it then suffices to bound $\mathsf{A}(P_i)/\mathsf{Opt}(P_i)$ for each part P_i.

Lemma 5.1.13. *At $1 + \mathcal{O}(\varepsilon)$ loss we can restrict to instances which consist of only one part.*

Proof. We consider an online algorithm A, some instance $\mathcal{I}$ and a partition of $\mathcal{I}$ into parts $P_i, i = 0, \ldots, \ell - 1$. We define an adapted version A' that applies $\mathsf{A}(P_i)$ to each part P_i and moves all jobs of P_i that are unfinished at the end of P_i to their respective safety nets. Recall that due to Lemma 5.1.9 these unfinished jobs must be released within the last period $Q_{a_{(i+1)}}$ of part P_i and this period is insignificant. Hence, these jobs contribute at most

$$\sum_{i=1}^{\ell+1} (1 + \varepsilon)^s rw(Q_{a_i}) \leq \sum_{i=1}^{\ell+1} 8\varepsilon \cdot \sum_{p=a_{i-1}+1}^{a_i-1} rw(Q_p) \leq \sum_{i=0}^{\ell} \mathcal{O}(\varepsilon) \cdot \mathsf{Opt}(P_i)$$

to the objective value. Therefore, we get that $\mathsf{A}'(\mathcal{I}) \leq (1 + \mathcal{O}(\varepsilon)) \sum_{i=0}^{\ell} \mathsf{A}(P_i)$. Applying $\sum_{i=0}^{\ell} \mathsf{Opt}(P_i) \leq \mathsf{Opt}(\mathcal{I})$ we can bound the competitive ratio of A' as follows:

$$\frac{\mathsf{A}'(\mathcal{I})}{\mathsf{Opt}(\mathcal{I})} \leq (1 + \mathcal{O}(\varepsilon)) \frac{\sum_{i=0}^{\ell} \mathsf{A}(P_i)}{\sum_{i=0}^{\ell} \mathsf{Opt}(P_i)} \leq (1 + \mathcal{O}(\varepsilon)) \max_{i=1,\ldots,\ell} \frac{\mathsf{A}(P_i)}{\mathsf{Opt}(P_i)}.$$

□

Note that there might be different partitions of one instance into parts since it is possible that a part is at the same time a sequence of significant periods. In the following section, we will consider online algorithms that work in principle only on relevant jobs while forgetting irrelevant jobs after they have been moved to their safety net. Hence, we need a partition into parts that can be determined regardless of the irrelevant jobs. To that end, we define $x(a) := (a+1)s - 1$ for a period Q_a to be the index of the last interval of period Q_a.

Lemma 5.1.14. *For a given instance $\mathcal{I}$, let $a_1 < ... < a_{\ell+1}$ be all indices such that*

$$(5.3) \quad (1+\varepsilon)^s \, rw(\mathrm{Rel}_{x(a_i)}(P_{i-1}) \cap Q_{a_i}) \leq$$

$$\varepsilon \cdot \left(1 + \frac{6\varepsilon}{(1+\varepsilon)^s}\right) \sum_{p=a_{i-1}+1}^{a_i-1} rw(\mathrm{Rel}_{x(a_i)}(P_{i-1}) \cap Q_p).$$

and $a_0 := -1$ where each P_{i-1} consists of all periods $Q_{a_{(i-1)}+1}, \ldots, Q_{a_i}$ for $i = 1, ..., \ell+1$. Then, each P_i is a part of $\mathcal{I}$ for $i = 0, ..., \ell$.

Proof. We first prove that the periods between Q_{a_i} and $Q_{a_{(i+1)}}$ build a sequence of significant periods. To that end, we can inductively apply Lemma 5.1.12 and get for each $a_{i-1} < a < a_i$:

$$(1+\varepsilon)^s \, rw(Q_a) \geq (1+\varepsilon)^s \, rw(\mathrm{Rel}_{x(a)}(P_{i-1}) \cap Q_a)$$

$$\overset{(5.3)}{>} \varepsilon \cdot \left(1 + \frac{6\varepsilon}{(1+\varepsilon)^s}\right) \sum_{p=a_{i-1}+1}^{a-1} rw(\mathrm{Rel}_{x(a)}(P_{i-1}) \cap Q_p)$$

$$= \varepsilon \cdot \left(1 + \frac{6\varepsilon}{(1+\varepsilon)^s}\right) rw\left(\mathrm{Rel}_{x(a)}(P_{i-1}) \cap \bigcup_{p=a_{i-1}+1}^{a-1} Q_p\right)$$

$$\overset{(5.2)}{\geq} \varepsilon \cdot \sum_{p=a_{i-1}+1}^{a-1} rw(Q_p).$$

That's why we can now apply Lemma 5.1.12 to each complete P_{i-1} and get that each Q_{a_i} is insignificant:

$$(1+\varepsilon)^s rw(Q_{a_i}) = (1+\varepsilon)^s rw(\mathrm{Rel}_{x(a_i)}(P_{i-1}) \cap Q_{a_i}) + (1+\varepsilon)^s rw(\mathrm{Ir}_{x(a_i)}(P_{i-1}) \cap Q_{a_i})$$

$$\leq \varepsilon \cdot \left(1 + \frac{6\varepsilon}{(1+\varepsilon)^s}\right) \cdot \sum_{p=a_{i-1}+1}^{a_i-1} rw(\mathrm{Rel}_{x(a_i)}(P_{i-1}) \cap Q_p)$$

$$+ 6\varepsilon \cdot rw(\mathrm{Rel}_{x(a_i)}(P_{i-1}))$$

$$\leq \varepsilon \cdot \left(1 + \frac{6\varepsilon}{(1+\varepsilon)^s} + 6\right) \cdot \sum_{p=a_{i-1}+1}^{a_i-1} rw(\mathrm{Rel}_{x(a_i)}(P_{i-1}) \cap Q_p)$$

$$+ 6\varepsilon \cdot rw(\mathrm{Rel}_{x(a_i)}(P_{i-1}) \cap Q_{a_i})$$

$$\leq \varepsilon \cdot \left(7 + \frac{6\varepsilon}{(1+\varepsilon)^s}\right) \cdot \sum_{p=a_{i-1}+1}^{a_i-1} rw(\mathrm{Rel}_{x(a_i)}(P_{i-1}) \cap Q_p)$$

$$+ \frac{6\varepsilon^2}{(1+\varepsilon)^s}\left(1 + \frac{6\varepsilon}{(1+\varepsilon)^s}\right) \cdot \sum_{p=a_{i-1}+1}^{a_i-1} rw(\mathrm{Rel}_{x(a_i)}(P_{i-1}) \cap Q_p)$$

$$\leq 8\varepsilon \cdot \sum_{p=a_{i-1}+1}^{a_i-1} rw(Q_p).$$

$\square$

We conclude this section by summarizing those consequences of our considerations that we need in the following section. To that end, we denote the maximum ratio between weights of relevant jobs at any point in time by $W = \frac{\Lambda \cdot \Gamma \cdot (1+\varepsilon)^{\Gamma+s}}{\varepsilon}$. Recall that the values of $\mu, \Lambda, s, K, \Gamma$, and hence W depend only on ε and m. Furthermore, for some given schedule we denote by $o_{x,j}$ the amount of a job j that is processed within an interval I_x.

Corollary 5.1.15. *At $1 + \mathcal{O}(\varepsilon)$-loss we can assume for each instance $\mathcal{I}$ with job set J and each interval I_x that*
(a) $p_j \in \{(1+\varepsilon)^k \mid \frac{\varepsilon^3}{4}(1+\varepsilon)^{-\Gamma} R_x \leq (1+\varepsilon)^k \leq \frac{1}{\varepsilon} R_x\}$ for each $j \in \mathrm{Rel}_x(J)$,
(b) $r_j \in \{(1+\varepsilon)^k \mid (1+\varepsilon)^{-\Gamma} R_x \leq (1+\varepsilon)^k \leq R_x\}$ for each $j \in \mathrm{Rel}_x(J)$,
(c) $w_j \in \{(1+\varepsilon)^k \mid w_x \leq (1+\varepsilon)^k \leq W \cdot w_x\}$ for some value w_x and each $j \in \mathrm{Rel}_x(J)$,
(d) $o_{x,j} \in \{\ell \cdot p_j/\mu \mid \ell \in \{0, \ldots, \mu\}\}$ for each $j \in \mathrm{Rel}_x(J)$,
(e) the cardinality of $\mathrm{Rel}_x(J)$ is bounded by $\Gamma\Lambda$, and
(f) $\sum_{j \in \mathrm{Ir}_x(J)} w_j C_j \leq O(\varepsilon) \cdot \mathrm{Opt}(\mathrm{Rel}_x(J))$.

Note that the respective sets of relevant processing times, release dates, weights and processing time fractions have constant size.

5.2 Abstraction of online algorithms

In this section we show how to construct a competitive-ratio approximation scheme based on the simplifications of Section 5.1. To do so, we restrict ourselves to such simplified instances and schedules. The key idea is to characterize the behavior of an online algorithm by a map: For each interval, the map gets as input the schedule computed so far and all information about the currently unfinished jobs. Based on this information, the map outputs how to schedule the available jobs within this interval. We denote the input by *configuration* and the output by *interval-schedule*. Confer Figure 5.1 for an illustration.

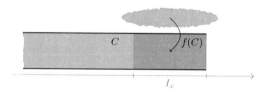

Figure 5.1: A configuration C for some interval I_x (in blue) consists of the information about all jobs released until time R_x and a schedule until I_x that was previously computed. Based on this information, an algorithm map f assigns an interval-schedule (in red) for the currently unfinished jobs to interval I_x.

Definition 5.2.1. An *interval-schedule* S for an interval I_x is defined by
– the index x of the interval,

– a set of jobs $J(S)$ available for processing in I_x together with the properties r_j, p_j, w_j of each job $j \in J(S)$ and its already finished part $o_j < p_j$ up to R_x,
– for each job $j \in J(S)$ the information whether j is relevant at time R_x, and
– for each job $j \in J(S)$ and each machine i a value q_{ij} specifying for how long j is processed by S on machine i during I_x.

An interval-schedule is called *feasible* if there is a feasible schedule in which all values for the jobs of $J(S)$ in the interval-schedule fit to the corresponding values of the schedule within the interval I_x. Denote the set of feasible interval-schedules as $\mathcal{S}$.

Definition 5.2.2. A *configuration* C for an interval I_x consists of
– the index x of the interval,
– a set of jobs $J(C)$ released up to time R_x together with the properties r_j, p_j, w_j, o_j of each job $j \in J(C)$,
– an interval-schedule for each interval $I_{x'}$ with $x' < x$.

A configuration is called *feasible* if there is a feasible schedule in which all values for the jobs of $J(C)$ in the configuration (including each interval-schedule) fit to the corresponding values of the schedule. The set of all feasible configurations (respecting the adaptations of Section 5.1) is denoted by $\mathcal{C}$. An *end-configuration* is a feasible configuration C for an interval I_x such that at time R_x, and not earlier, all jobs have been released and all relevant jobs have completely finished processing.

We say that an interval-schedule S is *feasible for a configuration* C if the set of jobs in $J(C)$ which are unfinished at time R_x matches the set $J(S)$ with respect to release dates, total and remaining processing time, weight and relevance of the jobs.

Instead of online algorithms we work from now on with *algorithm maps*, which are defined as functions $f : \mathcal{C} \to \mathcal{S}$. An algorithm map determines a schedule $f(\mathcal{I})$ for a given scheduling instance $\mathcal{I}$ by iteratively applying f to the corresponding configurations. W.l.o.g. we consider only algorithm maps f such that $f(C)$ is feasible for each configuration C and $f(\mathcal{I})$ is feasible for each instance $\mathcal{I}$. Call these algorithm maps *feasible*. Like for online algorithms, we define the competitive ratio ρ_f of an algorithm map f by $\rho_f := \max_{\mathcal{I}} f(\mathcal{I})/\mathsf{Opt}(\mathcal{I})$. Due to the following observation, algorithm maps are a natural generalization of online algorithms.

Proposition 5.2.3. *For each online algorithm* A *there is an algorithm map* f_A *such that when* A *is in configuration* $C \in \mathcal{C}$ *at the beginning of an interval* I_x, *algorithm* A *schedules the jobs according to* $f_A(C)$.

Recall that we restrict our attention to algorithm maps describing online algorithms which obey the simplifications introduced in Section 5.1. The essence of such online algorithms are the decisions for the relevant jobs. To this end, we define equivalence classes for configurations and for interval-schedules. Intuitively, two configurations (or interval-schedules) are equivalent if we can obtain one from the other by scalar multiplication with the same value, while ignoring the irrelevant jobs. The fact that in particular all old jobs are irrelevant induces that equivalence focuses on the constantly many latest configuration intervals, confer the illustration of Figure 5.2.

Definition 5.2.4. Let S, S' be two feasible interval-schedules for two intervals $I_x, I_{x'}$. Let further $\sigma : \tilde{J} \to \tilde{J}'$ be a bijection from a subset $\tilde{J} \subseteq J(S)$ to a subset $\tilde{J}' \subseteq J(S')$ and y an integer. The interval-schedules S, S' are (σ, y)-*equivalent* if $r_{\sigma(j)} = r_j(1 + \varepsilon)^{x'-x}, p_{\sigma(j)} =$

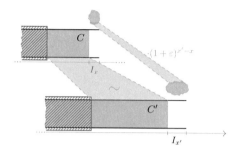

Figure 5.2: Illustration of two equivalent configurations C, C' where the relevant part of C' is a copy of the relevant part of C scaled in time by factor $(1+\varepsilon)^{x'-x}$.

$p_j(1+\varepsilon)^{x'-x}, o_{\sigma(j)} = o_j(1+\varepsilon)^{x'-x}, q_{i\sigma(j)} = q_{ij}(1+\varepsilon)^{x'-x}$ and $w_{\sigma(j)} = w_j(1+\varepsilon)^y$ for all $j \in \tilde{J}$ and $i = 1, \ldots, m$. Denote by $J_{\mathrm{Rel}}(S) \subseteq J(S)$ and $J_{\mathrm{Rel}}(S') \subseteq J(S')$ the jobs of $J(S)$ relevant at time R_x and of $J(S')$ relevant at time $R_{x'}$. The interval-schedules S, S' are *equivalent* (denoted by $S \sim S'$) if a bijection $\sigma : J_{\mathrm{Rel}}(S) \to J_{\mathrm{Rel}}(S')$ and an integer y exist such that they are (σ, y)-equivalent.

Definition 5.2.5. Let C, C' be two feasible configurations for two intervals $I_x, I_{x'}$. Denote by $J_{\mathrm{Rel}}(C), J_{\mathrm{Rel}}(C')$ the jobs which are relevant at times $R_x, R_{x'}$ in C, C', respectively. The configurations C, C' are *equivalent* (denoted by $C \sim C'$) if there is a bijection $\sigma : J_{\mathrm{Rel}}(C) \to J_{\mathrm{Rel}}(C')$ and an integer y such that
- $r_{\sigma(j)} = r_j(1+\varepsilon)^{x'-x}, p_{\sigma(j)} = p_j(1+\varepsilon)^{x'-x}, o_{\sigma(j)} = o_j(1+\varepsilon)^{x'-x}$ and $w_{\sigma(j)} = w_j(1+\varepsilon)^y$ for all $j \in J_{\mathrm{Rel}}(C)$, and
- the interval-schedules of I_{x-k} and $I_{x'-k}$ are (σ, y)-equivalent for each $k \in \mathbb{N}$.

The restriction of equivalence to relevant jobs allows a reasonable measurement of the performance of end-configurations contained in the same equivalence class. On the one hand we get equal performance ratios for equivalent configurations. On the other hand we can approximate the actual competitive ratio of the complete solution since relevant jobs dominate the objective value. Consider therefore an end-configuration C. We denote the objective value of a subset $\tilde{J} \subseteq J(C)$ in the history of C by $val_C(\tilde{J})$. We further define $\rho(C) := val_C(J_{\mathrm{Rel}}(C))/\mathrm{Opt}(J_{\mathrm{Rel}}(C))$ to be the achieved competitive ratio of C when restricted to the relevant jobs.

Lemma 5.2.6. *For each end-configuration $C \in \mathcal{C}$ it holds that*
(a) $(1+\mathcal{O}(\varepsilon))^{-1}\rho(C) \leq val_C(J(C))/\mathrm{Opt}(J(C)) \leq (1+\mathcal{O}(\varepsilon)) \cdot \rho(C)$ and
(b) $\rho(C) = \rho(C')$ for any $C' \in \mathcal{C}$ with $C \sim C'$.

Proof. The first property follows by Lemma 5.1.12 via
$$\frac{val_C(J(C))}{\mathrm{Opt}(J(C))} \leq \frac{val_C(J(C))}{\mathrm{Opt}(J_{\mathrm{Rel}}(C))} \leq (1+\mathcal{O}(\varepsilon))\frac{val_C(J_{\mathrm{Rel}}(C))}{\mathrm{Opt}(J_{\mathrm{Rel}}(C))} = (1+\mathcal{O}(\varepsilon))\rho(C).$$
Similarly, we get
$$\rho(C) = \frac{val_C(J_{\mathrm{Rel}}(C))}{\mathrm{Opt}(J_{\mathrm{Rel}}(C))} \leq \frac{val_C(J(C))}{\mathrm{Opt}(J_{\mathrm{Rel}}(C))} \leq (1+\mathcal{O}(\varepsilon))\frac{val_C(J(C))}{\mathrm{Opt}(J(C))}.$$

For the second property, denote the job weights and the resulting approximate interval bounds of the completion times of C (C') by w_j (w'_j) and D_j (D'_j). Since $C \sim C'$ there is an integer y and a bijection $\sigma : J_{\mathrm{Rel}}(C) \to J_{\mathrm{Rel}}(C')$ such that

$$
\begin{aligned}
val_{C'}(J_{\mathrm{Rel}}(C')) &= \sum_{j \in J_{\mathrm{Rel}}(C')} w'_j D'_j = \sum_{j \in J_{\mathrm{Rel}}(C)} w'_{\sigma(j)} D'_{\sigma(j)} \\
&= \sum_{j \in J_{\mathrm{Rel}}(C)} (1+\varepsilon)^y \, w_j \, (1+\varepsilon)^{x'-x} D_j \\
&= (1+\varepsilon)^{y+x'-x} \, val_C(J_{\mathrm{Rel}}(C)).
\end{aligned}
$$

If we transform the solution $\mathsf{Opt}(J_{\mathrm{Rel}}(C))$ by scaling weights by a factor of $(1+\varepsilon)^y$ and all time values by a factor of $(1+\varepsilon)^{x'-x}$ via σ we get a solution for $J_{\mathrm{Rel}}(C')$ with value $(1+\varepsilon)^{y+x'-x} \mathsf{Opt}(J_{\mathrm{Rel}}(C))$. Vice versa, applying the inverse transformation to the solution $\mathsf{Opt}(J_{\mathrm{Rel}}(C'))$ yields a solution for $J_{\mathrm{Rel}}(C)$ with value $(1+\varepsilon)^{x-x'-y} \mathsf{Opt}(J_{\mathrm{Rel}}(C'))$. By

$$
\begin{aligned}
\mathsf{Opt}(J_{\mathrm{Rel}}(C')) &\le (1+\varepsilon)^{y+x'-x} \mathsf{Opt}(J_{\mathrm{Rel}}(C)) \\
&\le (1+\varepsilon)^{y+x'-x} (1+\varepsilon)^{x-x'-y} \mathsf{Opt}(J_{\mathrm{Rel}}(C')) = \mathsf{Opt}(J_{\mathrm{Rel}}(C'))
\end{aligned}
$$

we can conclude equality. This finally yields

$$
\rho(C') = \frac{val_{C'}(J_{\mathrm{Rel}}(C'))}{\mathsf{Opt}(J_{\mathrm{Rel}}(C'))} = \frac{(1+\varepsilon)^{y+x'-x} \, val_C(J_{\mathrm{Rel}}(C))}{(1+\varepsilon)^{y+x'-x} \mathsf{Opt}(J_{\mathrm{Rel}}(C))} = \rho(C).
$$

$\square$

The following lemma shows that we can restrict the set of algorithm maps under consideration to those which treat equivalent configurations equivalently. We call algorithm maps obeying this condition (in addition to the restrictions of Section 5.1) *simplified algorithm maps*, see Figure 5.3. A configuration C is called *realistic for an algorithm map f* if there is an instance $\mathcal{I}$ such that if f processes $\mathcal{I}$ then at time R_x it is in configuration C.

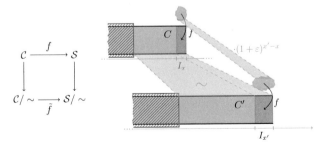

Figure 5.3: Since equivalent configurations are mapped to equivalent interval-schedules we get a commutative diagram for simplified algorithm maps.

Lemma 5.2.7. *At $1 + O(\varepsilon)$ loss we can restrict to algorithm maps f such that $f(C) \sim f(C')$ for any two equivalent configurations C, C'.*

Proof. Let f be an algorithm map. We now construct a new algorithm map $\bar{f}$ which is simplified and has competitive ratio $\rho_{\bar{f}} \leq (1 + \mathcal{O}(\varepsilon))\rho_f$ almost as good as f. Therefore, we pick for each equivalence class $\mathcal{C}_e \in \mathcal{C}/_\sim$ of the set of configurations a representative C_e (i.e. $[C_e] = \mathcal{C}_e$) which is realistic for f. For each configuration $C \in [C_e]$ which is equivalent to C_e with bijection σ and integer y, we define $\bar{f}$ by setting $\bar{f}(C)$ to be the interval-schedule for C which is (σ, y)-equivalent to $f(C_e)$. One can show by induction that $\bar{f}$ is always in a configuration such that an equivalent configuration is realistic for f. Hence, equivalence classes without realistic configurations for f are not relevant.

For an instance $\bar{\mathcal{I}}$ we show that there is an instance $\mathcal{I}$ such that $\bar{f}(\bar{\mathcal{I}})/\mathsf{Opt}(\bar{\mathcal{I}}) \leq (1 + \mathcal{O}(\varepsilon))f(\mathcal{I})/\mathsf{Opt}(\mathcal{I})$ which implies the claimed competitive ratio for $\bar{f}$. Let $\bar{C}$ for interval $I_{\bar{x}}$ be the end-configuration obtained when $\bar{f}$ is applied iteratively on $\bar{\mathcal{I}}$. Let C_e be the representative of the equivalence class of $\bar{C}$, which was chosen above and which is realistic for f (C_e is also an end-configuration). Therefore, there is an instance $\mathcal{I}$ such that C_e is reached at time R_x when f is applied on $\mathcal{I}$. Hence, by Lemma 5.2.6 and $C_e \sim \bar{C}$ we get that $\mathcal{I}$ is the required instance. □

Lemma 5.2.8. *There are only constantly many simplified algorithm maps. Each simplified algorithm map can be described using finite information.*

Proof. We assume the simplifications introduced in Section 5.1 and apply Corollary 5.1.15. Therefore, the domain of the algorithm maps under consideration contains only constantly many equivalence classes of configurations. Also, the target space contains only constantly many equivalence classes of interval-schedules. For an algorithm map f which obeys the restrictions of Section 5.1, the interval-schedule $f(C)$ is fully specified when knowing only C and the equivalence class which contains $f(C)$ (since the irrelevant jobs are moved to their safety net anyway). Since $f(C) \sim f(C')$ for a simplified algorithm map f if $C \sim C'$, we conclude that there are only constantly many simplified algorithm maps. Finally, each equivalence class of configurations and interval-schedules can be characterized using only finite information, and hence the same holds for each simplified algorithm map. □

The next lemma shows that up to a factor $1 + \varepsilon$, worst case instances of simplified algorithm maps span only constantly many intervals. Using this property, we will show in the subsequent lemmas that the competitive ratio of a simplified algorithm map can be determined algorithmically up to a $1 + \varepsilon$ factor.

Lemma 5.2.9. *There is a constant E such that for any instance I and any simplified algorithm map f there is a realistic end-configuration $\tilde{C}$ for an interval $I_{\tilde{x}}$ with $\tilde{x} \leq E$ which is equivalent to the corresponding end-configuration when f is applied to I.*

Proof. Consider a simplified algorithm map f. For each interval I_x, denote by $\mathcal{C}_x^f$ the set of realistic equivalence classes for I_x, i.e., the equivalence classes which have a realistic representative for I_x. Since there are constantly many equivalence classes and thus constantly many *sets* of equivalence classes, there must be a constant E independent of f such that $\mathcal{C}_{\bar{x}}^f = \mathcal{C}_{\bar{x}'}^f$ for some $\bar{x} < \bar{x}' \leq E$. Since f is simplified it can be shown by induction that $\mathcal{C}_{\bar{x}+k}^f = \mathcal{C}_{\bar{x}'+k}^f$ for any $k \in \mathbb{N}$, i.e., f *cycles* with period length $\bar{x}' - \bar{x}$.

Consider now some instance $\mathcal{I}$ and let C with interval I_x be the corresponding end-configuration when f is applied to $\mathcal{I}$. If $x \leq E$ we are done. Otherwise there must be some $k \leq \bar{x}' - \bar{x}$ such that $C^f_{\bar{x}+k} = C^f_{\bar{x}}$ since f cycles with this period length. Hence, by definition of $C^f_{\bar{x}+k}$ there must be a realistic end-configuration $\tilde{C}$ which is equivalent to C for the interval $I_{\tilde{x}}$ with $\tilde{x} := \bar{x} + k \leq E$. $\qquad\square$

Lemma 5.2.10. *Let f be a simplified algorithm map. There is an algorithm which approximates ρ_f up to a factor $1+\varepsilon$, i.e., it computes a value ρ' with $\rho' \leq \rho_f \leq (1+O(\varepsilon))\rho'$.*

Proof. According to Lemma 5.2.6 and Lemma 5.2.9, it is sufficient to construct the sets $C^f_0, ..., C^f_E$ in order to approximate the competitive ratio of all end-configurations in these sets. Due to Corollary 5.1.15, we know all possible values for all parameters of jobs explicitly with lower and upper bounds. Due to Lemma 5.1.14 we can additionally ensure that each equivalence class of configurations corresponds to at most one part. So the enumeration can be done in a finite amount of time. We start with C^f_0 and determine $f(C_e)$ for one representative C_e of each equivalence class $[C_e] \in C^f_0$. Based on this we determine the set C^f_1. We continue inductively to construct all sets C^f_x with $x \leq E$.

We define ρ_{max} to be the maximum ratio $\rho(C)$ for an end-configuration $C \in \cup_{0 \leq x \leq E} C^f_x$. Due to Lemma 5.2.9 and Lemma 5.2.6 the value ρ_{max} implies the required ρ' fulfilling the properties claimed in this lemma. $\qquad\square$

Our main algorithm works as follows. We first enumerate all simplified algorithm maps. For each simplified algorithm map f we approximate ρ_f using Lemma 5.2.10. We output the map f with the minimum (approximated) competitive ratio. Note that the resulting online algorithm has polynomial running time: All simplifications of a given instance can be done efficiently and for a given configuration, the equivalence class of the schedule for the next interval can be found in a look-up table of constant size.

Theorem 5.2.11. *$Pm | r_j, pmtn | \sum w_j C_j$ admits for any $m \in \mathbb{N}$ a competitive-ratio approximation scheme.*

5.3 Extension to non-preemptive scheduling

Certain arguments in Section 5.1 do not transfer directly to more complex scheduling settings. In this section, we argue how to overcome the increased complexity. If we are able to obtain similar statements as for Corollary 5.1.8 and Corollary 5.1.15 for the extended settings competitive-ratio approximation schemes are also possible.

In this section we review which statements or proofs of Section 5.1 make use of the possibility to preempt jobs and explain how to proceed in the non-preemptive case. One difference is that a schedule can be defined by a start time S_j for each job j with a resulting completion time of $C_j = S_j + p_j$ on its assigned machine i_j. Nevertheless, Lemma 5.1.1 can be proven similarly. We adapt the definition of time-stretch: when applying a time-stretch we now shift each completion time $C_j \in I_x$ to the next interval while keeping the offset wrt. the beginning of the interval, i.e., $C'_j = R_{x+1} + (C_j - R_x)$. For two completion time values $C_1 < C_2$ with $C_1 \in I_{x_1}$ and $C_2 \in I_{x_2}$ we observe that the difference after a time-stretch is $(C'_2 - C'_1) = (C_2 - C_1) + \sum_{x_1 \leq x < x_2} \varepsilon |I_x|$. Hence, idle time is inserted right before each job that is partially processed at some interval bound. Therefore, we adapt the proof of Lemma 5.1.2 as follows: for each interval I_x the large jobs with start- and completion time in I_x are shifted on each machine to the left such that on each machine the necessary amount of processing of small jobs can be assigned without preemption via

Smith's rule. Lemmas 5.1.4, 5.1.5, and 5.1.6 of Section 5.1.1 are not affected and remain valid as they are stated. Since splitting of jobs into small atoms as in Lemma 5.1.7 is not applicable here we give a variant concerning start times. For each job j we define $s(j)$ and $c(j)$ to be interval indices such that $S_j \in I_{s(j)}, C_j \in I_{c(j)}$. Recall that with our adapted objective function each ordering of jobs with $s(j) = c(j)$ assigned to one machine yields the same objective value. Hence, we only have to take care of the exact start times of those jobs with $s(j) < c(j)$ since they determine the amount of processing time assigned to each interval.

Lemma 5.3.1. *There is a constant $\mu \in \mathbb{N}$ such that at $1 + \varepsilon$ loss we can restrict to schedules where each large job j with $s(j) < c(j)$ has a start time of the form $S_j = R_{s(j)} + \ell_j \cdot |I_{s(j)}|/\mu$ with $\ell_j \in \{0, \ldots, \mu - 1\}$.*

Proof. We apply one time-stretch on a considered schedule yielding a new start time S_j for each job $j \in J$. Choose $\mu \in \mathbb{N}$ to be a constant integer such that $1/\mu$ is smaller than ε^2.

Consider now each interval I_x. For each machine i there is at most one job j with $x = s(j) < c(j)$ running on i. Due to the time-stretch there is idle time of at least $(1+\varepsilon)|I_{x-1}|$ before the start of j. We now set $\ell_j = \lfloor (S_j - R_x)\mu/|I_x| \rfloor$ and decrease the start time of j to $S'_j = R_x + \ell_j |I_x|/\mu$. This yields

$$S'_j \geq R_x + ((S_j - R_x)\frac{\mu}{|I_x|} - 1)\frac{|I_x|}{\mu} = S_j - \frac{|I_x|}{\mu} \geq S_j - \varepsilon^2 |I_x| \geq S_j - \varepsilon |I_{x-1}|.$$

Hence, we get a feasible schedule of the required form with costs increased by at most a factor of $(1 + \varepsilon)$. □

Therefore, a similar variant of Corollary 5.1.8 with a restatement of item (e) also holds in the non-preemptive setting.

The conclusion that also in the non-preemptive case only a constant amount of history is relevant demands a bit more work. First, the definition of the safety net (Lemma 5.1.9) needs to be adjusted since it might be that all machines are executing jobs during the entire interval I_{x+s-1}. However, with the adapted definition of time-stretching we know that there is reserved space on one machine in $[R_x, R_{x+s})$ to process all jobs released at time R_x. It remains to verify that an online algorithm (that does not know the future) can determine the beginning of this reserved space before it actually happens. Let j be the job scheduled to cover R_{x+s} on the last empty machine at this time. Denote its start and completion times before the time-stretch by $S_j \in I_{s(j)}, C_j \in I_{c(j)}$. From the above observation on time-stretches we can additionally conclude that $C'_j - p_j - \sum_{s(j) \leq x < c(j)} \varepsilon |I_x| \geq R_{s(j)+1}$. Hence, the safety net can be scheduled within interval $I_{s(j)+1}$ by our online algorithm since it decides about the complete schedule of an interval at its beginning.

With the existence of the safety net we can consider similar periods and the proof of Lemma 5.1.10 that only counts release weights of periods remains valid. For the following we need the small adaptation of $\Gamma := (K+1) \cdot s$. Definition 5.1.11 defining that a released job is recent, old, dominated, irrelevant and relevant at time R_x then uses this adapted Γ value. Nevertheless, the proof of Lemma 5.1.12 still holds. This is unfortunately not true for Lemma 5.1.13. Since some of the remaining jobs at the end of a part may have already started processing, we cannot simply move them to their safety net. Hence, parts cannot be treated independently. Therefore, we switch back to consider complete instances. If we can no longer assume that an instance is mainly a sequence of significant periods we lose the premise of Lemma 5.1.10. We cannot simply use that only a constant amount of history is relevant.

To solve this problem we will consider only special instances, where the weights after each insignificant period are sufficiently high such that they dominate the complete past until this period. The following lemma states the exact condition for these instances and proves again for each point in time that the relevant history of an instance with constant length dominates the irrelevant past even in the presence of insignificant periods. In a second step, we will reason why it is sufficient to consider only those instances satisfying the given condition.

To state the first lemma, we need the following definitions. For a given instance $\mathcal{I}$, let $a_1 < \cdots < a_{\ell+1}$ be again all indices satisfying Condition (5.3) of Lemma 5.1.14 (and $a_0 := -1$). Recall that the periods between $Q_{a_{i-1}}$ and Q_{a_i} build a sequence of significant periods for each $i = 1, \ldots, \ell$. Let first$(i), i = 1, \ldots, \ell$, denote the job that is released first after period Q_{a_i} at release time R_{x_i}. W.l.o.g. we can assume that $I_{x_i} \in Q_{a_i+1}$ since we can otherwise split the instance again after an empty period. For a given algorithm A we denote by $A(\mathcal{I}|i)$ the contribution of the jobs in $P_i = \bigcup_{p=a_i+1}^{a_{(i+1)}} Q_p$ to the objective value of the solution $A(\mathcal{I})$. We denote by $\mathcal{I}(i) := \cup_{k \leq i} P_k$ the instance up to period $Q_{a_{(i+1)}}$. ($I_{x_{(i+1)}}$ denotes the first interval after $Q_{a_{(i+1)}}$.)

Since we increased Γ by the length of one period and since $I_{x_i} \in Q_{a_i+1}$ we can use Lemma 5.1.10 to extend Lemma 5.1.12 such that $(1+\varepsilon)^s \cdot rw(\text{Ir}_x(P_{i-1})) \leq 6\varepsilon \cdot rw(\text{Rel}_x(P_{i-1}))$ actually holds for each $x_{(i-1)} \leq x < x_i, i = 1, \ldots, \ell+1$.

Lemma 5.3.2. *For an instance $\mathcal{I}$ with*

$$(5.4a) \qquad rw(\text{Rel}_{x_i-1}(\mathcal{I}(i-1))) \leq \frac{\varepsilon}{(1+\varepsilon)^s (1+7\varepsilon)} \cdot rw(\text{first}(i)) \quad and$$

$$(5.4b) \qquad \max\{w_j \mid j \in \text{Rel}_{x_i-1}(\mathcal{I}(i-1))\} \leq w_{\text{first}(i)}$$

for each $i = 1, \ldots, \ell$ it holds that

$$(5.5) \qquad (1+\varepsilon)^s rw(\text{Ir}_x(\mathcal{I}(i-1))) \leq 7\varepsilon \cdot rw(\text{Rel}_x(\mathcal{I}(i-1)))$$

for each $i = 1, \ldots, \ell+1$ and each $x_{i-1} \leq x < x_i$.

Proof. We proof this by induction and start with the base case $i = 1$. Consider some $x_0 \leq x < x_1$. By definition all periods before Q_{a_1} are significant. By Lemma 5.1.12 we get

$$(1+\varepsilon)^s rw(\text{Ir}_x(\mathcal{I}(0))) \leq 7\varepsilon \cdot rw(\text{Rel}_x(\mathcal{I}(0))).$$

For the inductive step consider some $i = 2, \ldots, \ell+1$ and assume that (5.5) holds for $i-1$ and each $x_{(i-2)} \leq x < x_{(i-1)}$. Observe that $\mathcal{I}(i-1) = \mathcal{I}(i-2) \cup P_{i-1}$. Due to Condition (5.4b) we know that $\text{Dom}_x(\mathcal{I}(i-1)) \cap P_{i-1} = \text{Dom}_x(P_{i-1})$ for each $x_{i-1} \leq x <$

x_i which yields:

$$
\begin{aligned}
(1+\varepsilon)^s \, rw(\mathrm{Ir}_x(\mathcal{I}(i-1))) \;=\; & (1+\varepsilon)^s \left[rw(\mathrm{Ir}_x(\mathcal{I}(i-1)) \cap \mathcal{I}(i-2)) \right. \\
& \left. + \, rw(\mathrm{Ir}_x(\mathcal{I}(i-1)) \cap P_{i-1}) \right] \\
\leq\; & (1+\varepsilon)^s \left[rw(\mathcal{I}(i-2)) + rw(\mathrm{Ir}_x(P_{i-1})) \right] \\
\overset{(5.5)}{\leq}\; & (1+\varepsilon)^s \left[(1+7\varepsilon) rw(\mathrm{Rel}_{x_{(i-1)}-1}(\mathcal{I}(i-2))) \right. \\
& \left. + \, rw(\mathrm{Ir}_x(P_{i-1})) \right] \\
\overset{(5.4)}{\leq}\; & \varepsilon \cdot rw(first(i-1)) + (1+\varepsilon)^s \cdot rw(\mathrm{Ir}_x(P_{i-1})) \\
\overset{(5.2)}{\leq}\; & \varepsilon \cdot rw(first(i-1)) + 6\varepsilon \cdot rw(\mathrm{Rel}_x(P_{i-1})) \\
\leq\; & 7\varepsilon \cdot rw(\mathrm{Rel}_x(\mathcal{I}(i-1))).
\end{aligned}
$$

This proves the hypothesis. □

Consider now some online algorithm A with competitive ratio ρ_{A} that performs with ratio $\overline{\rho_{\mathsf{A}}} := \max\{\mathsf{A}(\mathcal{I})/\mathsf{Opt}(\mathcal{I}) \mid \mathcal{I} \text{ satisfies } (5.4)\} \leq \rho_{\mathsf{A}}$ on our desired instances where the first released job after each insignificant period dominates the complete instance up to this period. We now want to find a new online algorithm A' that modifies the weights of each given arbitrary instance $\mathcal{I}$ to an instance $\mathcal{I}'$ satisfying (5.4) and creates a solution $\mathsf{A}(\mathcal{I}')$ that yields a schedule $\mathsf{A}'(\mathcal{I})$ for the original instance with

$$
\frac{\mathsf{A}'(\mathcal{I})}{\mathsf{Opt}(\mathcal{I})} \leq (1+\mathcal{O}(\varepsilon)) \max_i \frac{\mathsf{A}(\mathcal{I}'(i))}{\mathsf{Opt}(\mathcal{I}'(i))} \leq (1+\mathcal{O}(\varepsilon))\overline{\rho_{\mathsf{A}}} \leq (1+\mathcal{O}(\varepsilon))\rho_{\mathsf{A}}.
$$

Therefore, an online algorithm that proves to be good on only these instances yields an online algorithm for general instances with almost the same competitive ratio. Hence, it is sufficient to consider only these instances for the enumeration of simplified algorithm maps.

Lemma 5.3.3. *At $(1+\mathcal{O}(\varepsilon))$ loss, we can restrict to instances satisfying* (5.4).

Proof. For a considered online algorithm A we define a new algorithm A' that creates for a given instance $\mathcal{I}$ the new instance $\mathcal{I}'$ and applies A on $\mathcal{I}'$ as follows:

At the beginning we set $\mathcal{I}'(0) = \mathcal{I}(0)$. After period Q'_{a_i} we add for each further $i = 1, \ldots, \ell$ and for each job $j \in P_i$ of instance $\mathcal{I}$ a new job j' to $\mathcal{I}'$ with equal processing time and release date but new weight $w_{j'} := v_i \cdot w_j$ such that $w_{j'} \leq w_{first(i)'}$ for each $j' \in \mathrm{Rel}_{x_i-1}(\mathcal{I}'(i-1))$ and

$$
rw(\mathrm{Rel}_{x_i-1}(\mathcal{I}'(i-1))) \leq \frac{\varepsilon}{(1+\varepsilon)^s (1+7\varepsilon)} \cdot rw(first(i)').
$$

Hence, $\mathcal{I}'$ satisfies (5.4). The schedules for $\mathsf{A}'(\mathcal{I}(i))$ are then defined by applying $\mathsf{A}(\mathcal{I}'(i))$.

We then can easily observe that $\mathsf{A}'(\mathcal{I}|i) \cdot v_i = \mathsf{A}(I'|i) \le \mathsf{A}(I'(i))$ and get by Lemma 5.3.2

$$
\begin{aligned}
\mathsf{Opt}(\mathcal{I}'(i)) \;&\le\; \mathsf{Opt}(\mathcal{I}|i) \cdot v_i + (1+\varepsilon)^s \, rw(\mathcal{I}'(i-1)) \\
&\overset{(5.5)}{\le} \mathsf{Opt}(\mathcal{I}|i) \cdot v_i + (1+\varepsilon)^s \, (1+7\varepsilon) rw(\mathrm{Rel}_{x_i-1}(\mathcal{I}'(i-1))) \\
&\overset{(5.4)}{\le} \mathsf{Opt}(\mathcal{I}|i) \cdot v_i + \varepsilon \cdot rw(\mathrm{first}(i)') \le (1+\varepsilon) v_i \cdot \mathsf{Opt}(\mathcal{I}|i).
\end{aligned}
$$

The combination finally yields

$$
\frac{\mathsf{A}'(\mathcal{I})}{\mathsf{Opt}(\mathcal{I})} \le \max_i \frac{\mathsf{A}'(\mathcal{I}|i)}{\mathsf{Opt}(\mathcal{I}|i)} \le \max_i \frac{(1+\varepsilon) v_i \cdot \mathsf{A}(\mathcal{I}'(i))}{\mathsf{Opt}(\mathcal{I}'(i)) \cdot v_i} \le (1+\varepsilon) \max_i \frac{\mathsf{A}(\mathcal{I}'(i))}{\mathsf{Opt}(\mathcal{I}'(i))}
$$

Hence, we get $\rho_{\mathsf{A}'} \le (1+\mathcal{O}(\varepsilon)) \overline{\rho_{\mathsf{A}}} \le (1+\mathcal{O}(\varepsilon)) \rho_{\mathsf{A}}$. □

Corollary 5.3.4. *At $1+\mathcal{O}(\varepsilon)$ loss we can assume for each instance $\mathcal{I}$ with job set J and each interval I_x that*
(a) $p_j \in \{(1+\varepsilon)^k \mid \frac{\varepsilon^3}{4}(1+\varepsilon)^{-\Gamma} R_x \le (1+\varepsilon)^k \le \frac{1}{\varepsilon} R_x\}$ for each $j \in \mathrm{Rec}_x(J)$,
(b) $r_j \in \{(1+\varepsilon)^k \mid (1+\varepsilon)^{-\Gamma} R_x \le (1+\varepsilon)^k \le R_x\}$ for each $j \in \mathrm{Rec}_x(J)$,
(c) $w_j \in \{(1+\varepsilon)^k \mid w_x \le (1+\varepsilon)^k \le W \cdot w_x\}$ for some value w_x and each $j \in \mathrm{Rel}_x(J)$,
(d) $s(j), c(j) \in \{x-\Gamma, \ldots, x+s\}$ for each $j \in \mathrm{Rec}_x(J)$ and
    $S_j \in \{R_{s(j)} + \ell_j \cdot |I_{s(j)}|/\mu \mid \ell_j \in \{0, \ldots, \mu-1\}\}$ if $s(j) < c(j)$,
(e) the cardinality of $\mathrm{Rec}_x(J) \supseteq \mathrm{Rel}_x(J)$ is bounded by $\Gamma\Lambda$, and
(f) $\sum_{j \in \mathrm{Ir}_x(J)} w_j C_j \le O(\varepsilon) \cdot \mathsf{Opt}(\mathrm{Rel}_x(J))$.

We now continue with adaptations for Section 5.2. Since schedules are defined in the non-preemptive case via assigned machines and start times we discard the values of o_j and q_{ij} from the definitions of interval-schedules and configurations (cf. Definitions 5.2.1 and 5.2.2). Instead, each interval-schedule S for interval I_x knows in addition to all properties of $J(S)$ for each already running job j with $s(j) < x \le c(j)$ the values S_j and i_j and assigns these two values to each job j scheduled to start within I_x (i.e. $s(j) = x$). Also each configuration C for interval I_x contains additionally the values S_j and i_j for each job $j \in P(C) = \{j \in J(C) \mid s(j) < x \le c(j)\}$ that is partially processed at time R_x.

In Definition 5.2.4 of two (σ, y)-equivalent interval-schedules S, S' we replace the o_j- and q_{ij}-conditions by: $i_{\sigma(j)} = i_j$ and $x - x' = s(j) - s(\sigma(j)) = c(j) - c(\sigma(j))$ for each $j \in \tilde{J}$ and $S_{\sigma(j)} = S_j (1+\varepsilon)^{x'-x}$ for each $j \in \tilde{J}$ with $s(j) < x \le c(j)$.

In the non-preemptive setting it is possible that jobs running at the beginning of an interval are also dominated and hence irrelevant at that time. Nevertheless, configurations with different dominated jobs partially processed at time R_x must be treated differently. To deal with this circumstance we have to extend the definition of equivalent configurations appropriately:

Definition 5.3.5. Two feasible configurations C, C' for the intervals $I_x, I_{x'}$ are called equivalent
– if they are equivalent in the sense of Definition 5.2.5 without the o_j-conditions
– and if there is a bijection $\psi : P(C) \to P(C')$ such that $r_{\psi(j)} = r_j (1+\varepsilon)^{x'-x}$, $p_{\psi(j)} = p_j (1+\varepsilon)^{x'-x}$, $i_{\sigma(j)} = i_j$, and $S_{\sigma(j)} = S_j (1+\varepsilon)^{x'-x}$ for each $j \in P(C)$.

This extension allows a similar proof of Lemma 5.2.7 since for any feasible algorithm-map f a (σ, y)-equivalent interval schedule of $f(C_e)$ for any configuration $C \in [C_e]$ equivalent to the representative C_e has enough idle time where the running dominated jobs of $P(C)$ can be feasibly continued (and since Lemma 5.2.6 still holds).

Note that each job of $P(C)$ is recent at time R_x since no job remains unfinished for more than one period. Corollary 5.3.4 explicitly takes care on which statement is valid not only for the set of relevant but also for the set of recent jobs at any point in time. Therefore, we can use Corollary 5.3.4 in the proof of Lemma 5.2.8 instead of Corollary 5.1.15 to conclude again that there are only constantly many simplified algorithm maps. With this, the implications for Lemma 5.2.9 and Lemma 5.2.10 apply similarly and we can construct a competitive-ratio approximation scheme as in Section 5.2.

Theorem 5.3.6. $\mathrm{Pm}|\,r_j\,|\sum w_j C_j$ *admits a competitive-ratio approximation scheme for any* $m \in \mathbb{N}$.

6

Approximation Schemes
for Bidirectional Scheduling

This final chapter seeks for almost best possible off- and online algorithms for bidirectional scheduling with respect to performance guarantee. To that end, we ask for approximation schemes and provide for the single segment case both, a polynomial time approximation scheme for the offline setting as well as a competitive-ratio approximation scheme for the online setting. Therefore, we generalize PTAS techniques of Afrati et al. [Afr+99] and the concepts of Chapter 5 from machine scheduling to the bidirectional case.

This chapter is based on joined work with Yann Disser and Nicole Megow, cf. [Dis+15] for the first section.

Chapter 3 considers the complexity of bidirectional scheduling in particular with the restriction of unit processing times and weights. Hardness results are complemented with efficient exact algorithms. For arbitrary processing times the problem becomes NP-hard since single machine scheduling is a special case, cf. Theorem 3.0.1. In Chapter 4 we obtain first algorithmic results with provable performance guarantee for bidirectional jobs with arbitrary processing times and weights. Motivated by the considerations in Chapter 5 we now continue with the development of approximation schemes in this setting with the purpose of almost optimal off- and online algorithms.

The first step for competitive-ratio approximation schemes is the simplification of possible inputs and schedules under consideration (Section 5.1.1). This step heavily exploits PTAS techniques of Afrati et al. [Afr+99] for offline machine scheduling. Therefore, the simplification techniques presented therein depict a good foundation for our considerations.

As already discussed in Chapter 4, not every algorithmic technique for machine scheduling can be adapted easily to work in the bidirectional setting even for a single segment. There is e.g. the question of a reasonable incorporation of the transit times into LP relaxations based on time-indexed formulations or on completion or mean busy times. In particular, algorithms that rely on preemptive relaxations and simple ordering rules such as SPT ordering or Smith's rule are not obviously applicable. That's why we do not consider the first PTAS variant in [Afr+99] for a single segment and unit weights. Instead, the presented algorithm for parallel identical machines presented in [Afr+99] is more suitable for our purpose. The authors argue that it is possible at small loss to solve the problem by an outer dynamic program that uses an inner enumeration algorithm for a subset of jobs with constant size. Since enumeration can naturally be applied for bidirectional scheduling we mainly have to argue that sufficient assumptions for the dynamic program are valid. The main issue is to account for the different roles of processing and transit times for the interaction of jobs in the same and different directions. This is tackled by identifying the appropriate time value or job subset to cover by the distinct proof statements. It succeeded by now entirely for uniform compatibilities on a single segment.

The achieved simplifications are in addition a vital requirement for the development of a competitive-ration approximation scheme. With these in mind, there are mainly technical issues to be solved when extending the approach of Chapter 5. In particular, we have to deal with the fact that all jobs run with a transit time that remains equal throughout the complete instance. But the main idea remains as for non-preemptive machine scheduling. This is again due to the fact that the approximation scheme is based on enumeration.

6.1 Polynomial time approximation scheme

In what follows, we present a polynomial time approximation scheme for bidirectional scheduling for uniform compatibilities on a single segment with transit time τ, based on the ideas of Afrati et al. [Afr+99]. To that end, we focus the reasoning on the case where all jobs have a conflict. if the reasoning for the contrary case differs, it is actually easier to argue and the case distinctions would unnecessarily complicate matters.

As in the previous chapter, we proof several lemmas that allow us to make assumptions at "$1+\mathcal{O}(\varepsilon)$ loss", meaning that we can modify any input instance and optimum schedule to adhere to these assumptions, such that the resulting schedule is within a factor polynomial in $1+\mathcal{O}(\varepsilon)$ of the optimum schedule for the original instance where $\varepsilon \to 0$. For presentation reasons concerning the following section, these simplifications follow the proof scheme of Section 5.1.1.

The first crucial step is to restrict to processing times and release dates of the form $(1 + \varepsilon)^x$ for some $x \in \mathbb{N}$ and $r_j \geq \varepsilon(p_j + \tau)$. We can show that any instance can be adapted to have these properties, without making the resulting schedule worse by a factor of more than $(1 + \varepsilon)$.

Lemma 6.1.1. *With $1 + \mathcal{O}(\varepsilon)$ loss we can assume that $r_j, p_j \in \{(1 + \varepsilon)^x \mid x \in \mathbb{N}\} \cup \{0\}$, $r_j \geq \varepsilon(p_j + \tau)$, and $r_j \geq 1$ for each $j \in J$.*

Proof. Increasing any value $v \in \mathbb{R}$ to the smallest power of $(1 + \varepsilon)$ not smaller than v yields a value with $(1 + \varepsilon)^x = (1 + \varepsilon)(1 + \varepsilon)^{x-1} < (1 + \varepsilon)v$. Hence, multiplying all start times of a schedule by $(1 + \varepsilon)$ gives a feasible schedule even when rounded up all nonzero processing times to the next power of $(1+\varepsilon)$. The total completion time does not increase by more than a factor of $(1 + \varepsilon)$.

By shifting the completion times of a schedule with adapted processing times by a factor of $(1 + \varepsilon)$, we obtain increased start times S'_j for each job j:

$$S'_j = (1 + \varepsilon)C_j - (p_j + \tau) \geq (1 + \varepsilon)S_j + \varepsilon p_j + \varepsilon \tau \geq \varepsilon(p_j + \tau).$$

Hence, by losing not more than a $(1 + \varepsilon)$-factor we may assume that all jobs have release dates of at least an ε fraction of their running time. Now, we can scale the instance by some power of $(1 + \varepsilon)$, such that the earliest release date is at least one (since jobs with $r_j = p_j = \tau = 0$ can be ignored).

Finally, multiplying again all start times of a schedule with adapted processing times and release dates by $(1 + \varepsilon)$ yields a feasible schedule even when rounded up all nonzero release dates to the next power of $(1 + \varepsilon)$. □

We define $R_x = (1 + \varepsilon)^x$ and partition the time horizon into intervals $I_x = [R_x, R_{x+1}]$ of length εR_x, such that every job is released at the beginning of an interval. Since jobs are not released too early, we conclude that the maximum number of intervals σ covered by the running time of a single job is constant.

Lemma 6.1.2. *Each job runs for at most* $\sigma := \lceil \log_{1+\varepsilon} \frac{1+\varepsilon}{\varepsilon} \rceil$ *intervals, i.e., a job starting in interval* I_x *is completed before the end of* $I_{x+\sigma}$.

Proof. We consider some job j and assume that j starts in I_x in some schedule. By Lemma 6.1.1 we get

$$|I_x| = \varepsilon R_x \geq \varepsilon r_j \geq \varepsilon^2(p_j + \tau).$$

Thus, the running time of j is bounded by $|I_x|/\varepsilon^2$. The constant upper bound of $1/\varepsilon^2$ for the number of used intervals can still be improved since the length of the next σ succeeding intervals with increasing size is sufficient to cover a length of $|I_x|/\varepsilon^2$. Using the fact that $\sum_{k=0}^n z^k = \frac{1-z^{n+1}}{1-z}$ we get

$$\sum_{i=0}^{\sigma} |I_{x+i}| = \sum_{i=0}^{\sigma}(R_{x+i+1} - R_{x+i}) = |I_x| \sum_{i=0}^{\sigma}(1 + \varepsilon)^i$$

$$= |I_x| \frac{1 - (1 + \varepsilon)^{\sigma+1}}{1 - (1 + \varepsilon)}$$

$$\geq |I_x| \frac{1 - \frac{1+\varepsilon}{\varepsilon}}{-\varepsilon} = |I_x| \frac{1 + \varepsilon - \varepsilon}{\varepsilon^2} = \frac{|I_x|}{\varepsilon^2}$$

□

This allows us to group intervals together in blocks $B_t = \{I_{t\sigma}, I_{t\sigma+1}, \ldots, I_{(t+1)\sigma-1}\}$ of σ intervals each, such that every job scheduled to start in block B_t will terminate before the end of the next block B_{t+1}. We develop a dynamic program using the fact that each block only interacts with the next block. To specify an interface for this interaction we introduce the notion of a *frontier*. A block *respects an incoming frontier* $F = (f_l, f_r)$ if no leftbound (rightbound) job scheduled to start in the block starts earlier than f_l (f_r). Similarly, a block *respects an outgoing frontier* $F = (f_l, f_r)$ if no leftbound or rightbound job scheduled to start in the block would interfere with a leftbound (rightbound) job starting at time f_l (f_r). The symmetrical structure of the uniform compatibility graph (K_{n_r, n_l} or $\emptyset$) allows us to use this simple interface.

We introduce a dynamic programming table with entries $T[t, F, U]$ that are designed to hold the minimum total weighted completion time of scheduling all jobs in $U \subseteq J$ to start in block B_t or earlier, such that B_t respects the outgoing frontier F. We define $C(t, F_1, F_2, V)$ to be the minimum total weighted completion time of scheduling all jobs in V to start in B_t with B_t respecting the incoming frontier F_1 and the outgoing frontier F_2 (and ∞ if this is impossible). We have the following recursive formula for the dynamic programming table:

$$T[t, F, U] = \min_{F', V \subseteq U} \{T[t - 1, F', U \setminus V] + C(t, F', F, V)\}.$$

To turn this into an efficient dynamic program, we need to limit the dependencies of each entry and show that $C(\cdot)$ can be computed efficiently. The number of blocks to be considered can be polynomially bounded by $\log D$, where $D = \max_j r_j + n \cdot (\max_j p_j + \tau)$ is an upper bound on the makespan. The following proposition yields that we only need to consider polynomially many other entries to compute $T[t, F, U]$ and that we only need to evaluate $C(\cdot)$ for job sets of constant size, which we can do in polynomial time by simple enumeration. It will be proven in the following sequence of lemmas.

Proposition 6.1.3. *There is a schedule with a total weighted completion time within a factor of $1 + \mathcal{O}(\varepsilon)$ of the optimum and with the following properties:*
(a) The number of jobs scheduled in each block is bounded by a constant.
(b) Every two consecutive blocks respect one of constantly many frontiers.

We make again, at $1 + \varepsilon$ loss, the assumption that each job j finishing within I_x contributes $w_j \cdot D_j$ to the objective function where $D_j := R_{x+1}$.

For time-stretching in the bidirectional setting, we distinguish between two possible variants. Either each start time or each completion time is shifted to the next interval while maintaining the same offset to the beginning of the interval. By this, the schedule remains feasible and the objective is increased by a factor of at most $(1 + \varepsilon)$. When applying (multiple) time-stretches we use the following observation to assess the created empty space between two jobs of distinct intervals:

Lemma 6.1.4. *Consider two distinct time values $T_1 < T_2$ with $T_1 \in I_{x(1)}$ and $T_2 \in I_{x(2)}$. Applying ℓ time-stretches on these values yields shifted values $T_1' < T_2'$ with*

$$(6.1) \qquad (T_2' - T_1') \geq (T_2 - T_1) + \Xi[x(1), x(2)],$$

where $\Xi[x(1), x(2)] := \sum_{x(1) \leq x < x(2)} \ell \varepsilon |I_x|$.

Proof.

$$\begin{aligned}
(T_2' - T_1') &= R_{x(2)+\ell} + (T_2 - R_{x(2)}) - [R_{x(1)+\ell} + (T_1 - R_{x(1)})] \\
&= ((1 + \varepsilon)^\ell - 1)R_{x(2)} + T_2 - ((1 + \varepsilon)^\ell - 1)R_{x(1)} - T_1 \\
&\geq (T_2 - T_1) + (1 + \ell\varepsilon - 1)(R_{x(2)} - R_{x(1)}) \\
&= (T_2 - T_1) + \ell\varepsilon \sum_{x(1) \leq x < x(2)} |I_x|
\end{aligned}$$

$\square$

We can now apply time-stretches to the start or completion times of all jobs and use the above observation to quantify and locate the additional space created in the schedule. To that end, let $s(j)$ and $c(j)$ denote for each job j the interval indices with $S_j \in I_{s(j)}$ and $C_j \in I_{c(j)}$. Table 6.1 summarizes the resulting gaps between jobs $j, k \in J$ with starting times $S_j < S_k$ depending on whether start or completion times are stretched and whether j and k are aligned or opposed. Intuitively, applying a time-stretch via start times on two succeeding jobs j and k without intermediate idle time inserts the created gap of j's starting interval (and all further intervals before the start of k), i.e., $\Xi[s(j), s(k)]$, after job j. If applied on the completion times of j and k a time-stretch inserts the idle time of all intervals where the end is covered by job k, i.e., $\Xi[c(j), c(k)]$, before job k.

time-stretch on		aligned jobs	opposed jobs
start times	(6.1)	$S'_k \geq S'_j + p_j + \Xi[s(j), s(k)]$	$S'_k \geq S'_j + p_j + \tau + \Xi[s(j), s(k)]$
	$\Rightarrow$	$C'_k \geq C'_j + p_k + \Xi[s(j), s(k)]$	$C'_k \geq C'_j + p_k + \tau + \Xi[s(j), s(k)]$
compl. times	(6.1)	$C''_k \geq C''_j + p_k + \Xi[c(j), c(k)]$	$C''_k \geq C''_j + p_k + \tau + \Xi[c(j), c(k)]$
	$\Rightarrow$	$S''_k \geq S''_j + p_j + \Xi[c(j), c(k)]$	$S''_k \geq S''_j + p_j + \tau + \Xi[c(j), c(k)]$

Table 6.1: Resulting increased distance between two jobs $j, k \in J$ with $S_j < S_k$ after ℓ time-stretches. The shifted start- and completion times after time-stretches via start times are denoted by $'$ and by $''$ after time-stretches via completion times. Each first line follows by (6.1) using that jobs did not overlap before the time-stretch. Each second equation simply applies that $C'_j = S'_j + p_j + \tau$ and $S''_j = C''_j - p_j - \tau$ for each job j.

To analyze the set of jobs released within each interval we partition them as follows. A job j released at R_x is called *small* if $p_j \leq \frac{\varepsilon^2}{4}|I_x|$ and *large* otherwise. With this, we partition for each direction $d \in \{r, l\}$ the jobs $J^d_x := \{j \in J^d \mid r_j = R_x\}$ released at R_x into the subsets $S^d_x = \{j \in J^d_x \mid j \text{ is small}\}$ and $L^d_x = \{j \in J^d_x \mid j \text{ is large}\}$. We will see that the arrangement of jobs of each S^d_x does not influence the remaining jobs too much such that we can assume a fixed order for each of these sets. To do so, we denote again the sum of processing times of a subset $J' \subseteq J$ as $p(J')$ and the union of small jobs released up to some point R_x with direction $d \in \{r, l\}$ by $S^d_{\leq x} = \bigcup_{x' \leq x} S^d_x$. We now want to assume that each such subset is scheduled according to Smith's rule, i.e., $S_{j_1} \leq S_{j_2}$ for any pair of jobs $j_1, j_2 \in J'$ with $w_{j_2}/p_{j_2} < w_{j_1}/p_{j_1}$.

Lemma 6.1.5. *With $1 + \mathcal{O}(\varepsilon)$ loss we can restrict to schedules such that for each $x \geq 0$ and each $d \in \{r, l\}$:*

(a) the processing of no small job contains a release date,

(b) jobs of $S^d_{\leq x}$ available at time R_x are scheduled according to Smith's rule, and

(c) $p(S^d_x) \leq |I_x|$.

Proof. To prove claim (a) we consider some schedule and apply a time-stretch via start times. Observe that no further crossing of a processing over a release date is produced for small jobs. If there was a release date $R_{s(j)+1}$ contained in the processing interval of a small job of $I_{s(j)}$ it is moved behind the processing since we get by Equation (6.1) that $R_{s(j)+2} - S'_j \geq R_{s(j)+1} - S_j + \varepsilon|I_{s(j)}|$ which gives an increase larger than the processing time of this job.

For a proof of claim (b) consider a schedule S where no processing of a small job contains a release date and apply one time-stretch via start times. This increases the objective value by at most a $1 + \varepsilon$ factor. Denote the resulting schedule as S'. To achieve the demanded properties, apply the following procedure for each direction $d \in \{r, l\}$. First, remove all small jobs from schedule S'. Now consider each interval $I_x, x = 0, 1, \ldots$. Denote by A_x the set of removed jobs from I_x. If jobs have been removed in I_x there are idle intervals where jobs in direction d can be scheduled. Denote the subset of $S^d_{\leq x}$ already scheduled in earlier intervals by $B_{<x}$ and order the subset $C_x := S^d_{\leq x} \setminus B_{<x}$ of unscheduled jobs according to

Smith's rule. Define for $t \in I_x$ by $p_t(A_x) := p(\{j \in A_x \mid S'_j \leq t\})$ the amount of processing time of jobs started before time t in S'. Now let $C_x(t)$ be the smallest Smith ratio subset of C_x such that $p(C_x(t)) \geq p_t(A_x)$ or $C_x(t) = C_x$. Iterate from the earliest created maximal empty interval to the latest and fill each interval $[t_1, t_2]$ according to Smith's rule such that the jobs of $p(C_x(t_2))$ start before t_2. Note that $p(C_x(t_2)) \leq p_{t_2}(A_x) + \frac{\varepsilon^2}{4}|I_x|$ since we consider only small jobs. To maintain feasibility we increase the start of the following jobs from $J \setminus C_x$, if necessary. (This decreases eventually the size of the following empty interval which is no problem). Nevertheless, the start time of no job from $J \setminus C_x$ is increased by more than $\frac{\varepsilon^2}{4}|I_x|$. Hence, their completion time is increased by less than a $1 + \varepsilon$ factor and the jobs starting after R_{x+1} are not affected. Note that no processing of the assigned small jobs $B_x := C_x(R_{x+1})$ contains R_{x+1}.

With the adapted objective we can conclude as in Lemma 5.1.2 that the total weighted completion time of small jobs over all has not been increased.

To prove claim (c) consider for each $x = 0, 1, \ldots$ the largest Smith ratio subset J'_x of S^d_x, such that $p(J'_x) \leq |I_x|$. By assumptions (b) and (a) we can be sure that all jobs of $S^d_x \setminus J'_x$ are not scheduled within I_x and thus, we can move their release dates to R_{x+1}. □

Once, we have a fixed order to schedule small jobs with same release date we are able to glue them to job packs of a certain minimum size. For this purpose we apply a further time-stretch to join the processing of jobs assigned to the same pack. This increases for each interval I_y the amount of processing per direction and each earlier interval I_x by at most the size of one job being small at time R_x. Recall Lemma 5.1.3 (page 88) stating that $\sum_{x<y} \varepsilon^2 |I_x| \leq \varepsilon |I_y|$.

Lemma 6.1.6. *With $1 + \mathcal{O}(\varepsilon)$ loss we can restrict to schedules such that for each $x \geq 0$ and each $d \in \{r, l\}$ the jobs of S^d_x are joined according to Smith's rule to unsplittable job packs with size of at most $\frac{\varepsilon^2}{4}|I_x|$ and at least $\frac{\varepsilon^2}{8}|I_x|$ each.*

Proof. Consider a schedule satisfying at $1 + \mathcal{O}(\varepsilon)$ loss the properties of Lemma 6.1.5 and apply one time-stretch via start times. We now apply the following procedure for each direction $d \in \{r, l\}$ and each $x = 0, 1, \ldots$. Recall that the jobs of S^d_x are scheduled according to Smith's rule. Let $T^d_x = \{j \in S^d_x \mid p_j < \frac{\varepsilon^2}{8}|I_x|\}$ be the subset of jobs being too small. Remove the jobs of T^d_x from the current schedule and join the jobs of T^d_x successively in non-increasing Smith ratio order to minimal job packs such that the processing times of each job pack sum up to at least $\frac{\varepsilon^2}{8}|I_x|$. (The processing time of the last pack is artificially increased if necessary.) We now reassign complete job packs to the empty intervals similarly to the procedure in the proof of Lemma 6.1.5. Hence, no start time of T^d_x has been increased and the start time of no job in $J \setminus T^d_x$ has been increased by more than $\frac{\varepsilon^2}{4}|I_x|$.

In total, the start time of no job starting in interval I_{y+1} has been increased by more than $2 \cdot \sum_{x<y} \frac{\varepsilon^2}{4}|I_x| + 2 \cdot \frac{\varepsilon^2}{4}|I_y| \leq \varepsilon |I_y|$ due to Lemma 5.1.3. By Lemma 6.1.4 (or Table 6.1) we can conclude that no job has been delayed to a later interval by the rearrangement. Note that properties (a) and (c) of Lemma 6.1.5 still hold whereas property (b) (ordering according to Smith's rule) remains true only within each S^d_x. □

Therefore, we can consider each job pack simply as one small job where the processing time and weight of this job is equal to the respective sum. Besides the scheduling restrictions for small jobs we can also bound the amount of large jobs released at the beginning of each interval.

Lemma 6.1.7. *With $1 + \mathcal{O}(\varepsilon)$ loss we can assume for each $x \geq 0$ and each $d \in \{r,l\}$ that:*
(a) the number of possible processing times in L_x^d is bounded by $5 \log_{(1+\varepsilon)} \frac{1}{\varepsilon}$, and
(b) the number of jobs per processing time in L_x^d is bounded by $\frac{4}{\varepsilon^2}$.

Proof. Consider some scheduling instance, some $d \in \{r,l\}$ and some $x \geq 0$. The processing time of the jobs in L_x^d are, by definition, at least $\frac{\varepsilon^3}{4}(1+\varepsilon)^x$. On the other hand, by Lemma 6.1.1, the processing times are at most $\frac{1}{\varepsilon}(1+\varepsilon)^x$. Let x_j be such that $p_j = (1+\varepsilon)^{x_j}$. We get

$$\frac{\varepsilon^3}{4} \leq \frac{(1+\varepsilon)^{x_j}}{(1+\varepsilon)^x} \leq \frac{1}{\varepsilon}$$
$$\implies \log_{(1+\varepsilon)} \frac{\varepsilon^3}{4} \leq x_j - x \leq \log_{(1+\varepsilon)} \frac{1}{\varepsilon}$$

The difference of these bounds is $4 \log_{(1+\varepsilon)} \frac{1}{\varepsilon} + \log_{(1+\varepsilon)} 4$ which gives a constant number of possible integer values for x_j and, hence, a constant number of possible processing times for each job in L_x^d. Let p one of these. Observe that at most $4/\varepsilon^2$ jobs per direction with processing time $p \geq \frac{\varepsilon^2}{4}|I_x|$ can be schedule within I_x. The costs do not increase if we assume that the remaining jobs that need to start after R_{x+1} are those with smallest Smith ratio. Hence, their release date can be increased appropriately. $\square$

Lemma 6.1.8. *With $1 + \mathcal{O}(\varepsilon)$ loss we can assume, that each job is finished within a constant number of intervals after its release.*

Proof. Consider the set of jobs J_x released at time R_x. By Lemma 6.1.1 the running time of each such job is at most R_x/ε. Therefore, applying Lemmas 6.1.5 and 6.1.7 we can bound the time needed to first schedule all jobs of one direction and afterward all jobs of the other direction:

$$\sum_{d \in \{r,l\}} \left[p(S_x^d) + p(L_x^d) + \tau \right] \leq 2 \left[\varepsilon(1+\varepsilon)^x \right.$$
$$\left. + \frac{4}{\varepsilon^2} \cdot \frac{1}{\varepsilon}(1+\varepsilon)^x \cdot 5 \log_{(1+\varepsilon)} \frac{1}{\varepsilon} \right]$$
$$= \varepsilon^2(1+\varepsilon)^x \cdot 2 \left[\frac{1}{\varepsilon} + \frac{20}{\varepsilon^5} \log_{(1+\varepsilon)} \frac{1}{\varepsilon} \right]$$
$$\leq \varepsilon^2(1+\varepsilon)^x(1+\varepsilon)^{\sigma'-1} = \varepsilon |I_{x+\sigma'-1}|,$$

where σ' is the smallest possible integer such that $2 \left[\frac{1}{\varepsilon} + \frac{20}{\varepsilon^5} \log_{(1+\varepsilon)} \frac{1}{\varepsilon} \right] \leq (1+\varepsilon)^{\sigma'-1}$. Note, that σ' is constant.

Applying one time-stretch on the start times creates idle time for each interval I_x somewhere after σ' intervals that is sufficient to host all unfinished jobs of J_x, cf. Lemma 6.1.4 and Table 6.1. If no job was running at time $R_{x+\sigma'}$ before the time-stretch this created idle time is now part of interval $I_{x+\sigma'}$. Otherwise let j be the latest of these jobs with start time $S_j \in I_{s(j)}$ and completion time $C_j \in I_{c(j)}$ before the time-stretch. Note that $s(j) \leq x + \sigma' - 1$ which induces $c(j) \leq x + \sigma' + \sigma - 1$ due to Lemma 6.1.2. By Lemma 6.1.4 we can be sure that after the time-stretch there is idle time of $\sum_{x=s(j)}^{c(j)-1} \varepsilon |I_k|$

before (1) the start of the next job after j and (2) the end of interval I_{x_c+1}. By definition of σ', this time is sufficient to first schedule all jobs of J_x in heading of j and then all remaining. This way, all jobs of J_x are scheduled before the end of interval $I_{x+\sigma'+\sigma}$. $\square$

We can now limit the interface of our dynamic program. This finally proves Proposition 6.1.3 and therefore the following theorem.

Theorem 6.1.9. *The bidirectional scheduling problem with uniform compatibilities on a single segment admits a PTAS.*

Proof. By Lemma 6.1.6 we may assume that small jobs in S_x^d have processing time at least $\varepsilon^2 |I_x|/8$. By Lemma 6.1.5, the total processing time of these jobs is at most $|I_x|$, and hence the number of jobs in S_x^d is bounded by a constant. The same is true for large jobs, by Lemma 6.1.7. Finally, together with Lemma 6.1.8, this implies that the number of jobs running during each interval is bounded by a constant.

For the second property, we apply one time-stretch on the completion times. Consider now the latest job j of each direction that starts within block B_t and is completed in interval $I_{c(j)}$ of the following block. By Lemma 6.1.4 (and Table 6.1) we know that there is idle time of at least $\varepsilon |I_{c(j)-2}|$ before the start of job j (or before the start of the earliest job aligned with j with completion time in $I_{c(j)}$ and start time in B_t. Hence, we can decrease the start time of these jobs such that the values C_j and $S_j + p_j$ fall below the next ε^2 fraction of $I_{c(j)}$, i.e., by an amount of at most $\varepsilon^2 |I_{c(j)}| \leq \varepsilon |I_{c(j)-2}|$. Hence, the first job starting in B_{t+1} (of each direction in case of compatibilities) can be scheduled at an ε^2 fraction of $I_{c(j)}$ without any further loss. Thus, we only need to consider $\frac{\sigma}{\varepsilon^2}$ possible frontier values per direction, or a total of $\left(\frac{\sigma}{\varepsilon^2}\right)^2$ possible frontiers. $\square$

6.2 Competitive ratio approximation scheme

Finally, we argue for the existence of a competitive ratio approximation scheme for uniform compatibilities on a single segment. Lots of the necessary ingredients are already present. Some remaining issues have to be solved. Hence, we iterate in the following over the proof scheme of Chapter 5 and fill the remaining gaps.

For the geometric rounding (Lemma 5.1.1) we can extend Lemma 6.1.1 such that also the transit times and weights are powers of $1 + \varepsilon$. In the proof, the increase of the transit time can be realized together with the rounding of the processing times and is covered by the loss of factor $1 + \varepsilon$. Rounding the weights induces a further loss of factor $1 + \varepsilon$. These properties are not necessary for the PTAS, but since the transit times and weights must be part of the configurations we want to restrict their variety as well.

Lemma 6.2.1. *With $1 + \mathcal{O}(\varepsilon)$ loss we can assume that $r_j, p_j, \tau, w_j \in \{(1 + \varepsilon)^x \mid x \in \mathbb{N}\} \cup \{0\}$, $r_j \geq \varepsilon(p_j + \tau)$, and $r_j \geq 1$ for each $j \in J$.*

For a scheduled job j we denote again that $S_j \in I_{s(j)}$ as well as $C_j \in I_{c(j)}$ and consider again at $1 + \varepsilon$ loss the approximative objective function $\sum_j R_{c(j)+1}$. We keep the classification into small and large jobs as in Section 6.1, i.e., a job released at R_x is *small* if $p_j \leq \frac{\varepsilon^2}{4}|I_x|$ and *large* otherwise. Recall the notations of $J_x^d := \{j \in J^d \mid r_j = R_x\}$, $S_x^d = \{j \in J_x^d \mid j \text{ is small}\}$, and $L_x^d = \{j \in J_x^d \mid j \text{ is large}\}$.

The ordering of small jobs (Lemma 5.1.2) is covered by Lemma 6.1.5. Lemma 5.1.4 considering job packs of minimum size corresponds to Lemma 6.1.6 which must be slightly adapted. Since we want to restrict the diversity of possible inputs for a competitive-ratio approximation scheme we additionally have to round the new processing times and weights of job packs to powers of $1 + \varepsilon$. To ensure that the rounded processing times of the constructed job packs remain small we halve the criteria for the minimum size.

Lemma 6.2.2. *At $1 + \mathcal{O}(\varepsilon)$ loss we can restrict to instances such that $p_j \geq \frac{\varepsilon^2}{16}|I_x|$ for each $j \in S_x^d, x \geq 0, d \in \{r, l\}$. In these instances, the number of distinct processing times of each set S_x^d is bounded from above by $\log_{1+\varepsilon} 4$.*

Proof. Consider a schedule satisfying at $1 + \mathcal{O}(\varepsilon)$ loss the properties of Lemma 6.1.5 and apply one time-stretch via start times. We now apply the following procedure for each direction $d \in \{r, l\}$ and each $x = 0, 1, \ldots$. Recall that the jobs of S_x^d are scheduled according to Smith's rule. Let $T_x^d = \{j \in S_x^d \mid p_j < \frac{\varepsilon^2}{16}|I_x|\}$ be the subset of jobs being too small. Remove the jobs of T_x^d from the current schedule and join the jobs of T_x^d successively in non-increasing Smith ratio order to minimal job packs such that the processing times of each job pack sum up to at least $\frac{\varepsilon^2}{16}|I_x|$. (The processing time of the last pack is artificially increased if necessary.) We now reassign complete job packs to the empty intervals similarly to the procedure in the proof of Lemma 6.1.5. Hence, no start time of T_x^d has been increased and the start time of no job in $J \setminus T_x^d$ has been increased by more than $\frac{\varepsilon^2}{8}|I_x|$. We assign to each created job pack a processing time and weight equal to the respective sums of the jobs joint to that pack.

In total, the start time of no job starting in interval I_{y+1} has been increased by more than $2 \cdot \sum_{x<y} \frac{\varepsilon^2}{8}|I_x| + 2 \cdot \frac{\varepsilon^2}{8}|I_y| \leq \varepsilon|I_y|$ due to Lemma 5.1.3. By Lemma 6.1.4 (or Table 6.1) we can conclude that no job has been delayed to a later interval by the rearrangement.

Finally, at $1 + \mathcal{O}(\varepsilon)$ loss we can ensure that the processing times and weights of the new jobs are powers of $1 + \varepsilon$. Note that each new job remains small also after the rounding. Consequently, the processing times of jobs in S_x^d are of the form $(1 + \varepsilon)^y$ within the following range:

$$\frac{e^3}{16} \cdot (1 + \varepsilon)^x \leq (1 + \varepsilon)^y \leq \frac{e^3}{4}(1 + \varepsilon)^x.$$

The number of integers y satisfying these inequalities is bounded from above by $\log_{1+\varepsilon} 4$. □

Large jobs can be handled similar to machine scheduling as in Lemmas 5.1.5 and 5.1.6, confer Lemma 6.1.7.

We now take care of the transit time. It takes a special role when constructing a competitive-ratio approximation scheme since the transit time is fix for each instance. Hence, compared to the interval sizes it is "decreasing" from configuration to configuration. Nevertheless, it must be part of the configurations such that we are able to enumerate over a variety of instances. Hence, we have to bound the number of possible values for transit times by a constant. Due to Lemma 6.2.1 jobs of an instance with transit time τ are not released before time $\varepsilon\tau$ which yields a multiple of R_x with constant factor as upper bound on the possible transit times for each configuration for interval I_x. To get a similar lower bound we proof that the exact value of the transit time is not important once we have the information that it is relatively small compared to the current interval size, i.e., smaller than $\frac{\varepsilon^2}{3}|I_x|$. To that end, we want to *pretend* that each job in I_x runs with transit time $\tau(x) := \max\{\tau, \frac{\varepsilon^2}{3}|I_x|\}$ instead of τ. Given some schedule, we define for each job $j \in J$ the index $\hat{c}(j) := \min\{x \mid S_{j_1} + p_{j_1} + \tau(x) \in I_x\}$ and the resulting approximated completion time $\hat{C}_j := S_j + p_j + \tau(\hat{c}(j))$. For each index $x \geq 0$ we define the job set $T_x = \{j \in J \mid \hat{c}(j) = x\}$. For each job j with $S_j + p_j \in I_x$ it holds that $c(j) = \hat{c}(j)$ if $\frac{\varepsilon^2}{3}|I_x| \leq \tau$ and that $c(j) \leq \hat{c}(j) \leq c(j) + 1$ otherwise.

Lemma 6.2.3. *Consider some interval I_x. At $1 + \mathcal{O}(\varepsilon)$ loss we can assume*
(a) that $S_{j_1} + p_{j_1} + \tau(x) \leq S_{j_2}$ for any two opposed jobs j_1, j_2 with $j_1 \in T_x$ and $S_{j_1} \leq S_{j_2}$ having a conflict, and
(b) that each job $j \in T_x$ contributes $w_j \cdot R_{x+1}$ to the objective value.

Proof. Consider some schedule for an instance with transit time τ. We apply one time-stretch on the completion times yielding a loss of $1 + \varepsilon$. Due to Lemma 6.1.4 or Table 6.1 we can now be sure that $\hat{c}(j) = c(j)$ after the time-stretch. Hence, the objective value can be approximated as stated by Claim (b).

It remains to implement the headway between any two opposed jobs with conflict as claimed by (a). Consider now some interval I_x. If $\tau \geq \frac{\varepsilon^2}{3}|I_x|$ we are done. Otherwise we apply the following. Let A_x be the set of jobs with $s(j) < c(j) = x$, B_x the set of jobs with $s(j) = c(j) = x$, and C_x the set of jobs with $s(j) = x < c(j)$. If B_x is empty the claim follows since there is idle time of at least $\varepsilon|I_{x-1}| \geq \frac{\varepsilon^2}{3}|I_x|$ between any two opposed jobs of A_x and C_x (if they are not empty). Otherwise we use the observation of Lemma 4.1.6 how to schedule a set of available jobs with minimum length. Let $t_1 := \max\{R_x; S_j + p_j \mid j \in A_x\}$ and $t_2 := \min\{S_j \mid j \in C_x\}$. Due to the time-stretch we know that $t_2 - t_1 - \sum_{j \in B_x} p_j \geq \varepsilon|I_{x-1}| \geq \varepsilon^2|I_x|$. We schedule all rightbound jobs of B_x with a headway of $\frac{\varepsilon^2}{3}|I_x|$ after t_1 and after a further headway of $\frac{\varepsilon^2}{3}|I_x|$ between the processing intervals all leftbound jobs of B_x ensuring a third headway after the end of the last processing of $\frac{\varepsilon^2}{3}|I_x|$ before t_2. □

To conclude, we can assume that opposed jobs do not overlap even if we pretend that they travel with the adapted transit time. Since this transit time function is non-decreasing we can also be sure that aligned jobs with the adapted transit time do not overlap.

The obtained simplifications are summarized by the following corollary. The constant upper bound on the number of released jobs per interval (joint for both directions) Λ is an integer of at most $2\lceil \frac{16}{\varepsilon^2} + \frac{20}{\varepsilon^2} \log_{1+\varepsilon} \frac{1}{\varepsilon}\rceil$.

Corollary 6.2.4. *At $1 + \mathcal{O}(\varepsilon)$ loss we can assume that for each interval I_x*
(a) each job $j \in T_x$ is running with
   transit time $\tau(x) \in \{(1 + \varepsilon)^k \mid \frac{\varepsilon^3}{3} R_x \leq (1 + \varepsilon)^k \leq \frac{1}{\varepsilon} R_x\} \cup \{\frac{\varepsilon^3}{3} R_x\}$,
(b) each job j released at time R_x has
   processing time $p_j \in \{(1 + \varepsilon)^k \mid \frac{\varepsilon^3}{16} R_x \leq (1 + \varepsilon)^k \leq \frac{1}{\varepsilon} R_x\}$,
(c) there are at most $6 \log_{1+\varepsilon} \frac{1}{\varepsilon}$ distinct processing time values of jobs released at R_x,
(d) at most Λ jobs are released at R_x, and
(e) each small job starting in I_x finishes processing in I_x.

By now, we are missing a simplified representation of bidirectional schedules joining almost similar schedules to the same representation. The goal is to reduce the number of possibilities for the equivalence classes of interval schedules and configurations. As bidirectional jobs are inherently non-preemptive we do not use a generalization of

Lemma 5.1.7 for this purpose. Also Lemma 5.3.1, the non-preemptive equivalent cannot be simply adapted to the bidirectional setting. There can be several aligned jobs with completion time in the same interval and start time in earlier intervals with processing times arbitrarily short compared to the considered intervals. Hence, the start times cannot be shifted to interval fractions of fixed relative length. Nevertheless we will use these interval fractions to define a sufficient interface for jobs interfering with more than one interval. Therefore, we define for some time value t the following two procedures rounding t to the nearest $1/\mu$ fraction above or below t for some given constant $\mu \in \mathbb{N}$: $\lfloor t \rfloor_\mu := \max\{R_x + \ell \cdot \frac{|I_x|}{\mu} \le t \mid x \text{ with } t \in I_x, \ell = 0, \ldots, \mu - 1\}$ and $\lceil t \rceil_\mu := \min\{R_x + \ell \cdot \frac{|I_x|}{\mu} \ge t \mid x \text{ with } t \in I_x, \ell = 0, \ldots, \mu\}$. For a given schedule and two incompatible jobs j_1, j_2 we furthermore define

$$\widehat{\Delta}(j_1, j_2) := \begin{cases} p_{j_1} & \text{if } j_1 \text{ and } j_2 \text{ are aligned} \\ p_{j_1} + \tau(\hat{c}(j_1)) & \text{if } j_1 \text{ and } j_2 \text{ are opposed.} \end{cases}$$

This value determines the headway that must be hold between start times of two jobs. In the opposed case, it depends on the time where the first job is scheduled since we adapted the transit time to be time-dependent.

Lemma 6.2.5. *There is a constant $\mu \in \mathbb{N}$ such that we can restrict at $1 + \mathcal{O}(\varepsilon)$ loss to schedules with*

$$\lceil S_{j_1} + \widehat{\Delta}(j_1, j_2) \rceil_\mu \le \lfloor S_{j_2} \rfloor_\mu$$

for any two jobs j_1, j_2 with $s(j_1) < s(j_2)$ being incompatible.

Proof. Choose $\mu \in \mathbb{N}$ to be a constant integer such that $1/\mu$ is smaller than ε^2. Consider a schedule according to Lemma 6.2.3. With $1 + \varepsilon$ loss, we apply one time-stretch on the start times yielding new start times S_j with corresponding indices $s(j)$. Consider now each interval I_x and each pair of jobs j_1, j_2 with $s(j_1) < x = s(j_2)$ and $S_{j_1} + \widehat{\Delta}(j_1, j_2) \in I_x$. Since j_1 and j_2 did not overlap before the time-stretch (wrt. artificial transit times) we can conclude by Lemma 6.1.4 that

$$S_{j_2} \ge S_{j_1} + \widehat{\Delta}(j_1, j_2) + \varepsilon |I_{s(j_1) - 2}|$$

$$\ge S_{j_1} + \widehat{\Delta}(j_1, j_2) + \frac{I_x}{\mu}$$

Hence, there is at least one $\ell \in \{0, \ldots, \mu\}$ such that $R_x + \ell \cdot \frac{|I_x|}{\mu} \in [S_{j_1} + \Delta(j_1, j_2), S_{j_2}]$ which induces the claim. □

We now want to characterize a simplified representation of a schedule that only remembers $1/\mu$ fractions bounding the start time of each job from below end the end of processing and running from above. Note that two schedules with equal rounded completion times, i.e., $\lceil \widehat{C}_j \rceil_\mu = \lceil \widehat{C}'_j \rceil_\mu$ for each $j \in J$ have the same approximated objective value, since each job completes in both schedules within the same interval. Since these rounded values can not be injective, we give additionally a partial order $\zeta \subset J \times J$ on incompatible jobs as defined by non-increasing start times, i.e., any two incompatible jobs $(j_1, j_2) \in \zeta$ if and only if $S_{j_1} < S_{j_2}$.

Lemma 6.2.6. *Consider given $1/\mu$ fractions $\lfloor S_j \rfloor_\mu$, $\lceil S_j + p_j \rceil_\mu$ and $\lceil \widehat{C}_j \rceil_\mu$ for each $j \in J$ together with a partial order ζ and assume that there is a feasible schedule S that is consistent with these parameters and respects Lemmas 6.2.3 and 6.2.5. Then, a corresponding schedule S' can be constructed in polynomial time such that*
(a) ζ is respected by the start times of S',
(b) S' is feasible wrt. Lemma 6.2.3,
(c) $\lfloor S'_j \rfloor_\mu = \lfloor S_j \rfloor_\mu$, $\lceil S'_j + p_j \rceil_\mu \leq \lceil S_j + p_j \rceil_\mu$ and $\lceil \widehat{C}'_j \rceil_\mu \leq \lceil \widehat{C}_j \rceil_\mu$ for each $j \in J$, and
(d) Lemma 6.2.5 is satisfied.

Proof. We iterate over the jobs in topological order wrt. ζ and define new start times $S'_{j_2} :=$ $\max\{\lfloor S_{j_2} \rfloor_\mu\} \cup \{S'_{j_1} + \widehat{\Delta}'(j_1, j_2) \mid (j_1, j_2) \in \zeta\}$. Therefore, no two start times of S' violate the given partial order ζ. The resulting schedule S' is feasible and respects Lemma 6.2.3 since the respective distances between any two incompatible jobs are satisfied also by construction. Since each job $j \in J$ gets the minimum possible start time S'_j respecting the given constraints we can conclude all inequalities of property (c). This finally yields that Lemma 6.2.5 is guaranteed since $\lceil S'_{j_1} + \widehat{\Delta}'(j_1, j_2) \rceil_\mu \leq \lceil S_{j_1} + \widehat{\Delta}(j_1, j_2) \rceil_\mu \leq \lfloor S_{j_2} \rfloor_\mu = \lfloor S'_{j_2} \rfloor_\mu$ for any two incompatible jobs j_1, j_2 with $s(j_1) < s(j_2)$. □

Moreover, we can be sure that given $1/\mu$ fractions with a partial order own no corresponding schedule if the suggested procedure fails. If it succeeds and all values are equal we have a corresponding schedule. Otherwise there can be a difference of at most one $1/\mu$ fraction of the respective start interval between the rounded end values of the two schedules and we created a schedule of almost the same objective value where the upper bound of an end value is simultaneously a lower bound on the next rounded start value for all relevant cases of crossing jobs in the sense of Lemma 6.2.5. Of course, there is also an exact representation for this schedule which is also covered by the enumeration of possible equivalence classes. Therefore, we conclude the following.

Corollary 6.2.7. *At $1 + \mathcal{O}(\varepsilon)$ loss we can define a schedule by $(\lfloor S_j \rfloor_\mu, \lceil S_j + p_j \rceil_\mu, \lceil \widehat{C}_j \rceil_\mu), j \in J$ and ζ with $\lceil S_{j_1} + \widehat{\Delta}(j_1, j_2) \rceil_\mu \leq \lfloor S_{j_2} \rfloor_\mu$ for any two jobs $(j_1, j_2) \in \zeta$ with $s(j_1) < s(j_2)$.*

We continue to bound the length of the relevant history. Lemma 5.1.9, i.e., the safety net is covered by Lemma 6.1.8. We denote the respective number of intervals again by s. Therefore, we can consider a partition into periods with similar properties as in Section 5.1.2. The release weight of a set of jobs remains a lower bound to its contribution to the optimal value of the instance. Again, Lemma 5.1.10 applies in the bidirectional case since it considers only quantities of release weights for a sequence of significant periods. We now continue to follow the proof scheme for non-preemptive scheduling as given in Section 5.3. Hence, we use again $\Gamma := (K + 1) \cdot s$ to define that a released job is recent, old, dominated, irrelevant and relevant at time R_x, cf. Definition 5.1.11. As again only weights of released jobs are considered by the statements of Lemmas 5.1.12 and 5.1.14 and no structure of schedules is used for the proofs they hold also in the bidirectional setting. Hence, the release weights of the irrelevant jobs are sufficiently small compared to the release weights of the relevant jobs and we can identify a partition of an instance into parts by considering only relevant jobs. Also Lemmas 5.3.2 and 5.3.3 simply apply since they do not use any structure on feasible schedules apart from the fact that the release weight is a lower bound for the weighted completion time in any schedule and that a safety net can be guaranteed.

Corollary 6.2.8. *At* $1 + \mathcal{O}(\varepsilon)$ *loss we can assume for each instance* $\mathcal{I}$ *with job set* J *and each interval* I_x *that*

(a) $p_j \in \{(1+\varepsilon)^k \mid \frac{\varepsilon^3}{16}(1+\varepsilon)^{-\Gamma} R_x \leq (1+\varepsilon)^k \leq \frac{1}{\varepsilon} R_x\}$ *for each* $j \in \mathrm{Rec}_x(J)$,

(b) $r_j \in \{(1+\varepsilon)^k \mid (1+\varepsilon)^{-\Gamma} R_x \leq (1+\varepsilon)^k \leq R_x\}$ *for each* $j \in \mathrm{Rec}_x(J)$,

(c) $w_j \in \{(1+\varepsilon)^k \mid w_x \leq (1+\varepsilon)^k \leq W \cdot w_x\}$ *for some value* w_x *and each* $j \in \mathrm{Rel}_x(J)$,

(d) *for each* $(x - \Gamma) \leq x' \leq x$ *it holds that each job* $j \in T_{x'}$ *is running with transit time* $\tau(x') \in \{(1+\varepsilon)^k \mid \frac{\varepsilon^3}{3} R_{x'} \leq (1+\varepsilon)^k \leq \frac{1}{\varepsilon} R_{x'}\} \cup \{\frac{\varepsilon^3}{3} R_{x'}\}$,

(e) $\lfloor S_j \rfloor_\mu, \lceil S_j + p_j \rceil_\mu, \lceil \widehat{C}_j \rceil_\mu \in \{R_{x'} + \ell \cdot \frac{|I_{x'}|}{\mu} \mid x' = x - \Gamma, \ldots, x + s \text{ and } \ell = 0, \ldots, \mu\}$ *for each* $j \in \mathrm{Rec}_x(J)$,

(f) *the cardinality of* $\mathrm{Rec}_x(J) \supseteq \mathrm{Rel}_x(J)$ *is bounded by* $\Gamma\Lambda$, *and*

(g) $\sum_{j \in \mathrm{Ir}_x(J)} w_j C_j \leq O(\varepsilon) \cdot \mathrm{Opt}(\mathrm{Rel}_x(J))$.

We continue now with the adaptation of the abstraction step as described in Section 5.2. We start by giving appropriate definitions for interval-schedules and configurations based on the non-preemptive machine scheduling case. Instead of a machine assignment they get the additional information on the direction of each job j (denoted by $d_j \in \{\mathrm{r}, \mathrm{l}\}$) and the current transit time estimate.

Definition 6.2.9. An *interval-schedule* S for an interval I_x is defined by
– the index x of the interval,
– the transit time $\tau(x)$,
– a set of jobs $J(S)$ released but not completed by time R_x together with the properties r_j, p_j, w_j, d_j for each job $j \in J(S)$
– for each job $j \in J(S)$ the information whether j is relevant at time R_x, and
– a start time S_j for each job $j \in J(S)$ with $s(j) \leq x \leq \hat{c}(j)$.

Definition 6.2.10. A *configuration* C for an interval I_x consists of
– the index x of the interval,
– the transit time $\tau(x)$,
– a set of jobs $J(C)$ released up to time R_x together with the properties r_j, p_j, w_j, d_j for each job $j \in J(C)$,
– a start time S_j for each job $j \in P(C) = \{j \in J(C) \mid s(j) < x \leq \hat{c}(j)\}$ that is partially processed at time R_x
– an interval-schedule for each interval $I_{x'}$ with $x' < x$.

Therefore, we can use the respective terminology of Section 5.2 and apply the concept of algorithm maps corresponding to Proposition 5.2.3. We continue to adapt Definitions 5.2.4 and 5.2.5 of equivalent interval-schedules and configurations for the bidirectional case. To that end we apply the representation of bidirectional schedules corresponding to Corollary 6.2.7.

Definition 6.2.11. Let S, S' be two feasible interval-schedules for two intervals $I_x, I_{x'}$ with transit times $\tau(x), \tau'(x')$. Let further $\sigma : \tilde{J} \to \tilde{J}'$ be a bijection from a subset $\tilde{J} \subseteq J(S)$ to a subset $\tilde{J}' \subseteq J(S')$ and y an integer. The interval-schedules S, S' are (σ, y)-*equivalent* if
– $d_{\sigma(j)} = d_j$,
– $r_{\sigma(j)} = r_j(1+\varepsilon)^{x'-x}$, $p_{\sigma(j)} = p_j(1+\varepsilon)^{x'-x}$, $w_{\sigma(j)} = w_j(1+\varepsilon)^y$,
– $\lfloor S_{\sigma(j)} \rfloor_\mu = \lfloor S_j \rfloor_\mu (1+\varepsilon)^{x'-x}$, $\lceil S_{\sigma(j)} + p_{\sigma(j)} \rceil_\mu = \lceil S_j + p_j \rceil_\mu (1+\varepsilon)^{x'-x}$, and $\lceil \widehat{C}_{\sigma(j)} \rceil_\mu = \lceil \widehat{C}_j \rceil_\mu (1+\varepsilon)^{x'-x}$

for all $j \in \bar{J}$, if $S_{\sigma(j_1)} < S_{\sigma(j_2)} \iff S_{j_1} < S_{j_2}$ for any two jobs $j_1, j_2 \in \bar{J}$, and if $\tau'(x') = \tau(x)(1+\varepsilon)^{x'-x}$.

Definition 6.2.12. Let C, C' be two feasible configurations for two intervals $I_x, I_{x'}$. Denote by $J_{\mathrm{Rel}}(C), J_{\mathrm{Rel}}(C')$ the jobs which are relevant at times $R_x, R_{x'}$ in C, C', respectively. The configurations C, C' are *equivalent* (denoted by $C \sim C'$) if $\tau'(x') = \tau(x)(1+\varepsilon)^{x'-x}$, if there is a bijection $\sigma : J_{\mathrm{Rel}}(C) \to J_{\mathrm{Rel}}(C')$ and an integer y such that
- $d_{\sigma(j)} = d_j$,
- $r_{\sigma(j)} = r_j(1+\varepsilon)^{x'-x}, p_{\sigma(j)} = p_j(1+\varepsilon)^{x'-x}$, and $w_{\sigma(j)} = w_j(1+\varepsilon)^y$ for each $j \in J_{\mathrm{Rel}}(C)$,
- the interval-schedules of I_{x-k} and $I_{x'-k}$ are (σ, y)-equivalent for each $k \in \mathbb{N}$,

and if there is a bijection $\psi : P(C) \to P(C')$ such that
- $d_{\psi(j)} = d_j, r_{\psi(j)} = r_j(1+\varepsilon)^{x'-x}, p_{\psi(j)} = p_j(1+\varepsilon)^{x'-x}, \lceil S_{\psi(j)} + p_{\psi(j)} \rceil_\mu = \lceil S_j + p_j \rceil_\mu (1+\varepsilon)^{x'-x}$, and $\lceil \widehat{C}_{\psi(j)} \rceil_\mu = \lceil \widehat{C}_j \rceil_\mu (1+\varepsilon)^{x'-x}$ for each $j \in P(C)$.

We first observe that the proof of Lemma 5.2.10 works similarly since Lemma 5.1.12 is valid and the rounded completion times preserve the corresponding completion time intervals. With the same reasoning as for the non-preemptive case, we can apply the proof of Lemma 5.2.7 to argue that we can restrict to simplified algorithm maps. Even if two equivalent configurations C, C' have dominated jobs running at the beginning of the interval, it is ensured by the additional equivalence condition for these jobs that they are blocking new jobs from starting for the same amount of time, cf. Lemma 6.2.5. Hence, a feasible interval-schedule $f(C)$ for C defines a feasible interval-schedule for C' by equivalence. By the safety net we can be sure again that each job of $P(C)$ is recent at time R_x. Therefore, we can apply Corollaries 6.2.7 and 6.2.8 to conclude that there are only constantly many equivalence classes of configurations and constantly many equivalence classes of interval-schedules using the fact that there are only constantly many possibilities for ζ on the set $J_{\mathrm{Rel}}(C)$ with constant cardinality. Hence, Lemmas 5.2.8, 5.2.9 and 5.2.10 can be applied.

Theorem 6.2.13. *The bidirectional scheduling problem with uniform compatibilities on a single segment admits a competitive-ratio approximation scheme.*

Conclusions

In this thesis we investigate the situation where bidirectional traffic through a sequence of bottleneck segments is scheduled. These considerations were raised by a project with the Waterways and Shipping Board Kiel-Holtenau and Brunsbüttel concerning the ship traffic on the Kiel Canal. In addition to its bidirectional operation, the Kiel Canal constitutes a typical example of an online application.

Based on the occurring complex optimization problem we present a new compact model representing specific properties of bidirectional traffic on paths in general. One purpose of the thesis is to provide theoretical insights on these bidirectional characteristics paired with the problem's online nature. To that end, we discuss the problem's natural relation to classical machine scheduling. Analyzing similarities and differences fosters the development of algorithms with provable performance bounds on the one hand and the identification of hardness inducing properties on the other hand.

Table 1 summarizes again the complexity results for bidirectional scheduling. Besides the hardness raised by varying processing times, that is also present in machine scheduling, we which additional properties induce an increase in complexity. The presented polynomial time solvable special cases indicate that difficulties do not originate from the transit time of its own. But paired with an unbounded number of segments or with custom compatibilities NP-hardness arises. However, we identify a characterization of compatibilities that are less tough.

	Number m of segments		
compatibilities	$m = 1$	m const.	m arbitrary
Identical jobs $p_{ij} = p$, $\tau_{ij} = \tau_i$			
none	polynomial [Thm. 3.3.1]	polynomial[3] [Thm. 3.3.2]	NP-hard[1] [Thm. 3.1.1]
const. # of types			
arbitrary	NP-hard[2] [Thm. 3.2.5]		NP-hard[1,2]
Different jobs $p_{ij} = p_j$, $\tau_{ij} = \tau_i$			
uniform	NP-hard [Len+77]		NP-hard[1]

Table 1: Repetition of Table 3.1 for a summary of total completion time complexity for bidirectional scheduling.

In addition to the above offline algorithms that are efficient and exact we present algorithms with bounded ratio between resulting objective values optimal (offline) costs.

---

[1] even if $p = 0$, $\tau_i = 1$
[2] even if $\tau_i = p = 1$
[3] only if $p = 1, \tau_i \leq$ const, $r_j, \tau_i \in \mathbb{N}$

Table 2 gives an overview of these achievements. There is a general 4-competitive online framework that is able to tackle any variant of the BSP. For the special cases of uniform compatibilities on one or two segments we can guarantee a polynomial running time if weights are irrelevant. Otherwise, the single segment case can be adapted with a small increase of the approximation factor. Consequently, these variants admit constant factor approximation algorithms that are additionally competitive online algorithms. For the unweighted case with equal processing times there is a $(1 + \sqrt{2})$-competitive online algorithm. The dynamic program for the respective offline case needs complete information and cannot be applied in the online setting. Moreover, due to the respective lower bound of $(1 + \sqrt{5})/2$ for any competitive ratio no online algorithm can perform equivalently. The general 4-competitive online algorithm is complemented by a lower bound of 2.

		online				offline
			polynomial			
arbitrary instances		4 [4.1.3]				
Uniform compatibilities:			$(p_j, w_j$ id.)	$(w_j$ id.)		
$m = 1$	CRAS [6.2.13]	4	2.42 [4.2.6]	4 [4.1.7]	$4 + \varepsilon$ [4.1.9]	PTAS [6.1.9]
$m = 2$		4		4 [4.1.13]		
LB comp. ratio		2 [4.1.1]	1.6 [4.2.1]	2	2	

Table 2: Overview on algorithmic results. All but the last column are valid for the online setting. At the same time, the four right-most columns refer to polynomial algorithms. The lower bound of 2 on possible competitive ratios was proven for the unweighted case on a single segment with varying processing times and therefore holds for general instances. For the restricted setting with identical jobs a smaller lower bound was proven. The case of uniform compatibilities on a single segment admits a competitive-ratio approximation scheme (CRAS).

Dealing with online scheduling problems with a proper gap between lower and upper bound on the optimal competitive ratio we introduce the concept of competitive-ratio approximation schemes. Such schemes compute online algorithms with a competitive ratio arbitrarily close to the best possible competitive ratio. To that end, we present techniques that provide a new and interesting view on the behavior of online algorithms.

We provide such schemes for online parallel machine scheduling of preemptive and non-preemptive jobs to minimize the weighted sum of completion times. Extensions for related and unrelated machines, arbitrary monomial cost functions, the makespan and for randomized online algorithms are possible [Gün+13]. In particular, recent results building on [Gün+13] show that the concept can also be applied to other scheduling settings [Kur+13], other online models [Che+15; MW13], and also to different online problems such as the k-server problem [Möm13]. We believe that our considerations have the capability to contribute to the understanding of online algorithms. Examples are the conclusions that almost best possible competitive ratios can be achieved by efficient online algorithms or that lower bound instances with constant length can get close to the actual worst-case examples.

Returning to the bidirectional setting, we show that the approach of competitive-ratio approximation schemes can be applied to the BSP on a single segment. The necessary simplification techniques yield at the same time that there is a PTAS for the respective offline variant. The question of possible generalizations of the simplification ideas to either a constant number of compatibility types (maybe even "velocity types") or a constant number of segments with bounded transit time ratio remains open.

A further interesting open issue remains from the computational point of view. After the discussions about a combinatorial relaxation of the STCP (Section 1.3.2) the question of MIP formulations with stronger LP relaxations arose. We considered two different approaches: a column generation model, where columns represent schedules on transit segments, and a time indexed formulation for train timetabling by Borndörfer and Schlechte [BS07; Sch11]. For the latter model we used an existing implementation [TSO11] of the authors after creating an appropriate space discretization and assuming an empty compatibility graph. First experiments for both models indicate that further intensive work is necessary on that issue.

For the original complex problem of ship traffic control at the Kiel Canal we integrated algorithmic ideas from two important related applications, train scheduling on a single-track network and collision-free routing of automated guided vehicles. This unified view of routing and scheduling allowed us to find a blend between simultaneous (global) and sequential (local) solution approaches to allot scarce network resources to a fleet of vehicles in a collision-free manner.

We developed a fast heuristic, with an average running time of less than two minutes, which yields solutions that are approved by the expert planners. Instance-dependent lower bounds prove the quality of the achieved objective function values which significantly improve upon manual plans. Much more importantly, we model the practical context in such a high level of detail that the resulting tool perfectly reflects the effects of enlargement options at the Kiel Canal. This enabled the officials to evaluate the different options under ship traffic predicted for the year 2025, and to base their decisions on the simulation results.

Even though it came as a side effect of the study, we lay the ground for a computer aided traffic control. In fact, our planning in rolling horizons integrates with a heuristic [Luy10] that schedules the locking process at each boundary of the Kiel Canal since entering, passing, and exiting the canal are interdependent. The overall system may support the expert planners during several potentially difficult years of construction work. Moreover, it was considered to use the tool for deciding about the schedule of the construction work itself: Different orders of construction and the selection of different excavating machines (several of which significantly hinder regular traffic) directly influence the traffic flow under scarce resources.

A next step for further interesting work would be theoretical and computational considerations of traffic in general bidirectional networks. In particular for our heuristic ideas respective extensions appear conceivable.

Bibliography

[Afr+99]   F. N. Afrati, E. Bampis, C. Chekuri, D. R. Karger, C. Kenyon, S. Khanna,
            I. Milis, M. Queyranne, M. Skutella, C. Stein, and M. Sviridenko. "Approx-
            imation Schemes for Minimizing Average Weighted Completion Time with
            Release Dates". In: *Proc. 40th Symposium on Foundations of Computer Sci-
            ence (FOCS)*. 1999, pp. 32–43.

[All+08]   A. Allahverdi, C. T. Ng, T. C. E. Cheng, and M. Y. Kovalyov. "A survey of
            scheduling problems with setup times or costs". In: *European J. Oper. Res.*
            187.3 (2008), pp. 985 –1032. DOI: `DOI:10.1016/j.ejor.2006.06.060`.

[AP02]   E. J. Anderson and C. N. Potts. "On-line Scheduling of a Single Machine to
            Minimize Total Weighted Completion Time". In: *Proc. 13th Annual ACM-
            SIAM Symposium on Discrete Algorithms (SODA)*. SIAM, 2002, pp. 548–
            557.

[AP04]   E. J. Anderson and C. N. Potts. "Online Scheduling of a Single Machine
            to Minimize Total Weighted Completion Time". In: *Math. Oper. Res.* 29.3
            (2004), pp. 686–697. DOI: `10.1287/moor.1040.0092`.

[Ant+14]   A. Antoniadis, N. Barcelo, D. Cole, K. Fox, B. Moseley, M. Nugent, and
            K. Pruhs. "Packet Forwarding Algorithms in a Line Network". In: *Proc.
            11th Latin American Theoretical Informatics Symposium (LATIN)*. Vol. 8392.
            LNCS. 2014, pp. 610–621. DOI: `10.1007/978-3-642-54423-1_53`.

[Aug+08]   J. Augustine, S. Irani, and C. Swamy. "Optimal Power-Down Strategies". In:
            SIAM J. Comput. 37.5 (Jan. 2008), pp. 1499–1516. DOI: `10.1137/05063787X`.

[Bak+80]   B. Baker, E. Coffman Jr., and R. Rivest. "Orthogonal Packings in Two Dimen-
            sions". In: *SIAM J. Comput.* 9.4 (1980), pp. 846–855. DOI: `10.1137/0209064`.

[BD+94]   S. Ben-David, A. Borodin, R. Karp, G. Tardos, and A. Wigderson. "On the
            power of randomization in on-line algorithms". In: *Algorithmica* 11.1 (1994),
            pp. 2–14. DOI: `10.1007/BF01294260`.

[Błą+13]   I. Błądek, M. Drozdowski, F. Guinand, and X. Schepler. *On contiguous and
            non-contiguous parallel task scheduling*. Research Report RA-6/2013. Insti-
            tute of Computing Science, Poznań University of Technology, 2013.

[BS07]   R. Borndörfer and T. Schlechte. "Models for Railway Track Allocation". In:
            *Proc. 7th Workshop on Algorithmic Approaches for Transportation Modelling,
            Optimization, and Systems (ATMOS)*. 2007. DOI: `10.4230/OASIcs.ATMOS.
            2007.1170`.

[BEY05]   A. Borodin and R. El-Yaniv. *Online Computation and Competitive Analysis*.
            Cambridge University Press, 2005.

[Bru+05]   P. Brucker, S. Knust, and G. Wang. "Complexity results for flow-shop problems with a single server". In: *European J. Oper. Res.* 165 (2 2005), pp. 398–407.

[CG94]   X. Cai and C. Goh. "A fast heuristic for the train scheduling problem". In: *Comput Oper Res* 21.5 (1994), pp. 499 –510. DOI: http://dx.doi.org/10.1016/0305-0548(94)90099-X.

[CL95]   M. Carey and D. Lockwood. "A Model, Algorithms and Strategy for Train Pathing". In: *J. Oper. Res. Soc.* 46 (1995), pp. 988–1005. DOI: 10.1057/jors.1995.136.

[CB73]   J. L. Carroll and M. S. Bronzini. "Waterway transportation simulation models: Development and application". In: *Water Resour. Res.* 9.1 (1973), pp. 51–63. DOI: 10.1029/WR009i001p00051.

[Ced07]   A. Ceder. *Public Transit Planning and Operation: Theory, Modelling and Practice.* Oxford: Butterworth-Heinemann (Elsevier), 2007.

[Cha+96]   S. Chakrabarti, C. Phillips, A. S. Schulz, D. B. Shmoys, C. Stein, and J. Wein. "Improved scheduling algorithms for minsum criteria". In: *Proc. 23rd International Conference on Automata, Languages and Programming (ICALP).* Vol. 1099. LNCS. 1996, pp. 646–657. DOI: 10.1007/3-540-61440-0_166.

[Che+01]   C. Chekuri, R. Motwani, B. Natarajan, and C. Stein. "Approximation Techniques for Average Completion Time Scheduling". In: *SIAM J. Comput.* 31 (2001), pp. 146–166. DOI: 10.1137/S0097539797327180.

[Che+15]   L. Chen, D. Ye, and G. Zhang. "Approximating the Optimal Algorithm for Online Scheduling Problems via Dynamic Programming". In: *Asia-Pacific Journal of Operational Research* 32.01 (2015), p. 1540011. DOI: 10.1142/S0217595915400114.

[Chu+10]   C. Chung, T. Nonner, and A. Souza. "SRPT is 1.86-Competitive for Completion Time Scheduling". In: *Proc. 21st Annual ACM-SIAM Symposium on Discrete Algorithms (SODA).* SIAM, 2010, pp. 1373–1388.

[Chv83]   V. Chvátal. *Linear programming.* Macmillan, 1983.

[CS11]   S. Coene and F. C. R. Spieksma. "The lockmasters' problem". In: *Proc. 11th Workshop on Algorithmic Approaches for Transportation Modelling, Optimization, and Systems (ATMOS).* 2011, pp. 27–37.

[Cor+09]   T. Cormen, C. Leiserson, R. Rivest, and C. Stein. *Introduction to Algorithms.* 3. MIT Press, 2009.

[CW09]   J. Correa and M. Wagner. "LP-Based Online Scheduling: From Single to Parallel Machines". In: *Math. Program.* 119 (1 2009). 10.1007/s10107-007-0204-7, pp. 109–136.

[Dak65]   R. J. Dakin. "A tree-search algorithm for mixed integer programming problems". In: *The Computer Journal* 8.3 (1965), pp. 250–255. DOI: 10.1093/comjnl/8.3.250.

[DS88]   M. Desrochers and F. Soumis. "A generalized permanent labeling algorithm for the shortest path problem with time windows". In: *INFOR: Information Systems and Operational Research* 26.3 (1988), pp. 191–212.

[Des+95]   J. Desrosiers, Y. Dumas, M. M. Solomon, and F. Soumis. "Time constrained routing and scheduling". In: *Network Routing*. Ed. by M. O. Ball, T. L. Magnanti, C. L. Monma, and G. L. Nemhauser. Vol. 8. Handbooks in Operations Research and Management Science. Elsevier, 1995. Chap. 2, pp. 35–139.

[Dij59]   E. W. Dijkstra. "A note on two problems in connexion with graphs". In: *Numer. Math.* 1 (1 1959), pp. 269–271.

[Dis+15]   Y. Disser, M. Klimm, and E. Lübbecke. "Scheduling Bidirectional Traffic on a Path". In: *Proc. 42nd Colloquium on Automata, Languages, and Programming (ICALP)*. Vol. 9134. LNCS. 2015, pp. 406–418. DOI: 10.1007/978-3-662-47672-7_33.

[DVDS06]   C. Duin and E. Van Der Sluis. "On the Complexity of Adjacent Resource Scheduling". In: *J. Sched.* 9.1 (2006), pp. 49–62. DOI: 10.1007/s10951-006-5593-6.

[ES11]   T. Ebenlendr and J. Sgall. "Semi-Online Preemptive Scheduling: One Algorithm for All Variants". In: *Theory Comput. Syst.* 48.3 (2011), pp. 577–613. DOI: 10.1007/s00224-010-9287-2.

[Ebe+09]   T. Ebenlendr, W. Jawor, and J. Sgall. "Preemptive Online Scheduling: Optimal Algorithms for All Speeds". In: *Algorithmica* 53 (2009), pp. 504–522. DOI: 10.1007/s00453-008-9235-6.

[ES03]   L. Epstein and R. van Stee. "Lower Bounds for On-line Single-Machine Scheduling". In: *Theoretical Computer Science* 299.1–3 (2003), pp. 439 –450. DOI: 10.1016/S0304-3975(02)00488-7.

[FW98a]   A. Fiat and G. J. Woeginger. "Competitive odds and ends". In: *Online Algorithms: The State of the Art*. Vol. 1442. LNCS. 1998, pp. 385–394. DOI: 10.1007/BFb0029578.

[FW98b]   A. Fiat and G. J. Woeginger, eds. *Online Algorithms: The State of the Art*. Vol. 1442. LNCS. Springer Berlin Heidelberg, 1998. DOI: 10.1007/BFb0029561.

[Fis+03]   A. V. Fishkin, K. Jansen, and M. Mastrolilli. "On Minimizing Average Weighted Completion Time: A PTAS for the Job Shop Problem with Release Dates". In: *Proc. 14th Symposium an Algorithms and Computation (ISAAC)*. Vol. 2906. LNCS. 2003, pp. 319–328. DOI: 10.1007/978-3-540-24587-2_34.

[FF58]   L. Ford and D. Fulkerson. "Constructing Maximal Dynamic Flows from Static Flows". In: *Oper. Res.* 6.3 (1958), pp. 419–433.

[FF62]   L. Ford and D. Fulkerson. *Flows in networks*. Princeton: Princeton University Press, 1962.

[Gaf+15]   E. R. Gafarov, A. Dolgui, and A. A. Lazarev. "Two-station single-track railway scheduling problem with trains of equal speed". In: *Computers & Industrial Engineering* 85.0 (2015), pp. 260 –267. DOI: http://dx.doi.org/10.1016/j.cie.2015.03.014.

[GJ79]   M. R. Garey and D. S. Johnson. *Computers and intractability: A Guide to the Theory of NP-Completeness*. W. H. Freeman & Co., New York, 1979.

[Gar+76]   M. R. Garey, D. S. Johnson, and R. Sethi. "The complexity of flowshop and jobshop scheduling". In: *Math. Oper. Res.* 1.2 (1976), pp. 117–129.

[Gar+73]   M. Garey, R. Graham, and J. Ullman. "An analysis of some packing algorithms". In: *Combinatorial algorithms (Courant Comput. Sci. Sympos., No.
9, 1972)* (1973), pp. 39–47.

[Gaw+08]   E. Gawrilow, E. Köhler, R. Möhring, and B. Stenzel. "Dynamic Routing of
Automated Guided Vehicles in Real-time". In: *Mathematics: Key Technology
for the Future*. Ed. by H.-J. Krebs and W. Jäger. Berlin, Heidelberg: Springer,
2008, pp. 165–177.

[Gho94]   J. B. Ghosh. "Batch scheduling to minimize total completion time". In: *Oper.
Res. Lett.* 16.5 (1994), pp. 271–275. DOI: DOI:10.1016/0167-6377(94)
90040-X.

[Goe97]   M. X. Goemans. "Improved approximation algorithms for scheduling with
release dates". In: *Proc. 8th Annual ACM-SIAM Symposium on Discrete Algorithms (SODA)*. SIAM, 1997, pp. 591–598.

[Goe+02]   M. X. Goemans, M. Queyranne, A. S. Schulz, M. Skutella, and Y. Wang.
"Single Machine Scheduling with Release Dates". In: *SIAM J. Discrete Math.*
15.2 (2002), pp. 165–192. DOI: 10.1137/S089548019936223X.

[Gra+79]   R. L. Graham, E. L. Lawler, J. K. Lenstra, and A. H. G. Rinnooy Kan. "Optimization and Approximation in deterministic sequencing and scheduling: A
survey". In: *Ann. Discrete Math.* 5 (1979), pp. 287–326.

[Grö+81]   M. Grötschel, L Lovász, and A. Schrijver. "The ellipsoid method and its
consequences in combinatorial optimization". In: *Combinatorica* 1.2 (1981),
pp. 169–197. DOI: 10.1007/BF02579273.

[Grö+01]   M. Grötschel, S. O. Krumke, J. Rambau, T. Winter, and U. T. Zimmermann.
"Combinatorial Online Optimization in Real Time". In: *Online Optimization
of Large Scale Systems*. Ed. by M. Grötschel, S. Krumke, and J. Rambau.
Springer Berlin Heidelberg, 2001, pp. 679–704. DOI: 10.1007/978-3-662-
04331-8_33.

[Gün+13]   E. Günther, O. Maurer, N. Megow, and A. Wiese. "A New Approach to
Online Scheduling: Approximating the Optimal Competitive Ratio". In: *Proc.
24th Annual ACM-SIAM Symposium on Discrete Algorithms (SODA)*. SIAM,
2013, pp. 118–128. DOI: 10.1137/1.9781611973105.9.

[Gün+14]   E. Günther, F. G. König, and N. Megow. "Scheduling and Packing Malleable
and Parallel Tasks with Precedence Constraints of Bounded Width". In: *J.
Comb. Optim.* 27.1 (2014), pp. 164–181. DOI: 10.1007/s10878-012-9498-3.

[Hal+96]   L. A. Hall, D. B. Shmoys, and J. Wein. "Scheduling to Minimize Average Com-
pletion Time: Off-Line and On-Line Approximation Algorithms". In: *Proc. 7th
Annual ACM-SIAM Symposium on Discrete Algorithms (SODA)*. Vol. 96.
SIAM, 1996, pp. 142–151.

[Hal+97]   L. A. Hall, A. S. Schulz, D. B. Shmoys, and J. Wein. "Scheduling to Minimize
Average Completion Time: Off-Line and On-Line Approximation Algorithms".
In: *Math. Oper. Res.* 22.3 (1997), pp. 513–544. DOI: 10.1287/moor.22.3.513.

[Hig+97] A. Higgins, E. Kozan, and L. Ferreira. "Heuristic Techniques for Single Line Train Scheduling". In: *J. Heuristics* 3.1 (1997), pp. 43–62. DOI: 10.1023/A: 1009672832658.

[HS87] D. S. Hochbaum and D. B. Shmoys. "Using Dual Approximation Algorithms for Scheduling Problems Theoretical and Practical Results". In: *J. ACM* 34.1 (Jan. 1987), pp. 144–162. DOI: 10.1145/7531.7535.

[Hoo+98] H. Hoogeveen, P. Schuurman, and G. J. Woeginger. "Non-approximability Results for Scheduling Problems with Minsum Criteria". In: *Proc. 6th International Conference on Integer Programming and Combinatorial Optimization (IPCO)*. Vol. 1412. LNCS. 1998, pp. 353–366. DOI: 10.1007/3-540-69346-7_27.

[HV96] J. Hoogeveen and A. Vestjens. "Optimal on-line algorithms for single-machine scheduling". In: *Proc. 5th International Conference on Integer Programming and Combinatorial Optimization (IPCO)*. Vol. 1084. LNCS. 1996, pp. 404–414. DOI: 10.1007/3-540-61310-2_30.

[IK75] O. H. Ibarra and C. E. Kim. "Fast Approximation Algorithms for the Knapsack and Sum of Subset Problems". In: *J. ACM* 22.4 (Oct. 1975), pp. 463–468. DOI: 10.1145/321906.321909.

[Joh73] D. S. Johnson. "Approximation Algorithms for Combinatorial Problems". In: *Proc. 5th Annual ACM Symposium on Theory of Computing (STOC)*. STOC '73. ACM, 1973, pp. 38–49. DOI: 10.1145/800125.804034.

[Kar+88] A. R. Karlin, M. S. Manasse, L. Rudolph, and D. D. Sleator. "Competitive snoopy caching". In: *Algorithmica* 3 (1988), pp. 79–119. DOI: 10.1007/BF01762111.

[Kar72] R. M. Karp. "Reducibility among Combinatorial Problems". In: *Complexity of Computer Computations*. Ed. by R. E. Miller, J. W. Thatcher, and J. D. Bohlinger. The IBM Research Symposia Series. 1972, pp. 85–103. DOI: 10.1007/978-1-4684-2001-2_9.

[Kel+04] H. Kellerer, U. Pferschy, and D. Pisinger. *Knapsack problems*. Springer Science & Business Media, 2004.

[Kie14] Kiel Canal Official Website. http://www.kiel-canal.org/english.htm. last accessed November 17, 2014. 2014.

[Koc+11] T. Koch, T. Achterberg, E. Andersen, O. Bastert, T. Berthold, R. Bixby, E. Danna, G. Gamrath, A. Gleixner, S. Heinz, A. Lodi, H. Mittelmann, T. Ralphs, D. Salvagnin, D. Steffy, and K. Wolter. "MIPLIB 2010 – Mixed Integer Programming Library version 5". In: *Math. Progr. Comp.* 3.2 (2011), pp. 103–163.

[TSO11] Konrad-Zuse-Zentrum für Informationstechnik Berlin (ZIB). *Train Schedule Optimization tool TS-OPT*. 2011.

[Kru+03] S. O. Krumke, W. E. de Paepe, D. Poensgen, and L. Stougie. "News from the online traveling repairman". In: *Theoretical Computer Science* 295.1–3 (2003), pp. 279 –294. DOI: 10.1016/S0304-3975(02)00409-7.

[Kur+13]   A. Kurpisz, M. Mastrolilli, and G. Stamoulis. "Competitive Ratio Approxima-
            tion Schemes for Makespan Scheduling Problems". In: *Proc. 10th Workshop
            on Approximation and Online Algorithms (WAOA)*. Vol. 7846. LNCS. 2013,
            pp. 159–172. DOI: 10.1007/978-3-642-38016-7_14.

[Lab+84]   J. Labetoulle, E. L. Lawler, J. K. Lenstra, and A. H. G. Rinnooy Kan. "Pre-
            emptive scheduling of uniform machines subject to release dates". In: *Progress
            in Combinatorial Optimization (Waterloo, Ont., 1982)*. Toronto, Ont.: Aca-
            demic Press Canada, 1984, pp. 245–261.

[LD60]   A. H. Land and A. G. Doig. "An Automatic Method of Solving Discrete
            Programming Problems". In: *Econometrica* 28.3 (1960), pp. 497–520.

[Law+93]   E. L. Lawler, J. K. Lenstra, A. H. G. Rinnooy Kan, and D. B. Shmoys.
            "Sequencing and Scheduling: Algorithms and Complexity". In: *Handbooks in
            Operations Research and Management Science*. Ed. by S. Graves, A. R. Kan,
            and P. Zipkin. Vol. 4. Amsterdam, The Netherlands: Elsevier, 1993, pp. 445–
            522.

[Lei+94]   F. Leighton, B. M. Maggs, and S. B. Rao. "Packet routing and job-shop
            scheduling in O(congestion+dilation) steps". In: *Combinatorica* 14.2 (1994),
            pp. 167–186.

[Len+77]   J. K. Lenstra, A. H. G. Rinnooy Kan, and P. Brucker. "Complexity of machine
            scheduling problems". In: *Ann. Discrete Math.* 1 (1977), pp. 343–362.

[Leu04]   J. Y.-T. Leung. *Handbook of scheduling: Algorithms, Models, and Perfor-
            mance Analysis*. Vol. 1. CRC Press, 2004.

[LL09]   P. Liu and X. Lu. "On-line scheduling of parallel machines to minimize to-
            tal completion times". In: *Computers and Operations Research* 36 (2009),
            pp. 2647–2652. DOI: 10.1016/j.cor.2008.11.008.

[Lu+03]   X. Lu, R. A. Sitters, and L. Stougie. "A class of on-line scheduling algorithms
            to minimize total completion time". In: *Oper. Res. Lett.* 31 (2003), pp. 232–
            236. DOI: 10.1016/S0167-6377(03)00016-6.

[Lüb+14]   E. Lübbecke, M. E. Lübbecke, and R. H. Möhring. *Ship Traffic Optimization
            for the Kiel Canal*. Tech. rep. 4681. Available at http://www.optimization-
            online.org/DB_HTML/2014/12/4681.html. Optimization Online, Dec. 2014.

[LR94]   C. Lund and N. Reingold. "Linear programs for randomized on-line algo-
            rithms". In: *Proc. 5th Annual ACM-SIAM Symposium on Discrete Algorithms
            (SODA)*. SIAM, 1994, pp. 382–391.

[Lus+11]   R. M. Lusby, J. Larsen, M. Ehrgott, and D. Ryan. "Railway track allocation:
            models and methods". In: *OR Spectrum* 33.4 (2011), pp. 843–883. DOI: 10.
            1007/s00291-009-0189-0.

[Luy10]   M. Luy. "Algorithmen zum Scheduling von Schleusungsvorgängen am Beispiel
            des Nord-Ostsee-Kanals". In German. Available at www.diplom.de/e-book/
            228321/algorithmen-zum-scheduling-von-schleusungsvorgaengen-am-
            beispiel-des-nord-ostsee-kanals. MA thesis. TU Berlin, Institut für
            Mathematik, 2010.

[Mao+95]   W. Mao, R. Kincaid, and A. Rifkin. "On-Line Algorithms for a Single Machine Scheduling Problem". In: *The Impact of Emerging Technologies on Computer Science and Operations Research*. Ed. by S. Nash, A. Sofer, W. Stewart, and E. Wasil. Vol. 4. Operations Research/Computer Science Interfaces Series. Springer US, 1995, pp. 157–173. DOI: 10.1007/978-1-4615-2223-2_8.

[McN59]   R. McNaughton. "Scheduling with Deadlines and Loss Functions". In: *Manage. Sci.* 6.1 (1959), pp. 1–12. DOI: 10.1287/mnsc.6.1.1.

[Meg07]   N. Megow. "Coping with incomplete information in scheduling—stochastic and online models". PhD thesis. TU Berlin, 2007.

[MS04]   N. Megow and A. S. Schulz. "On-line scheduling to minimize average completion time revisited". In: *Oper. Res. Lett.* 32.5 (2004), pp. 485–490. DOI: 10.1016/j.orl.2003.11.008.

[MW13]   N. Megow and A. Wiese. "Competitive-Ratio Approximation Schemes for Minimizing the Makespan in the Online-List Model". In: *CoRR* abs/1303.1912 (2013). Available at http://arxiv.org/abs/1303.1912.

[Mit+13]   K. N. Mitchell, B. X. Wang, and M. Khodakarami. "Selection of Dredging Projects for Maximizing Waterway System Performance". In: *Transp. Res. Rec.* 2330 (2013), pp. 39–46. DOI: 10.3141/2330-06.

[Möm13]   T. Mömke. "A Competitive Ratio Approximation Scheme for the k-Server Problem in Fixed Finite Metrics". In: *CoRR* abs/1303.2963 (2013). Available at http://arxiv.org/abs/1303.2963.

[Ng+03]   C. T. Ng, T. C. E. Cheng, J. J. Yuan, and Z. H. Liu. "On the single machine serial batching scheduling problem to minimize total completion time with precedence constraints, release dates and identical processing times". In: *Oper. Res. Lett.* 31.4 (2003), pp. 323–326. DOI: DOI:10.1016/S0167-6377(03)00007-5.

[Pas+14]   W. Passchyn, D. Briskorn, and F. C. R. Spieksma. "Mathematical programming models for scheduling locks in sequence". In: *Proc. 14th Workshop on Algorithmic Approaches for Transportation Modelling, Optimization, and Systems (ATMOS)*. Vol. 42. Schloss Dagstuhl–Leibniz-Zentrum fuer Informatik. 2014, pp. 92–106.

[PW11]   B. Peis and A. Wiese. "Universal packet routing with arbitrary bandwidths and transit times". In: *Proc. 15th International Conference on Integer Programming and Combinatorial Optimization (IPCO)*. Vol. 6655. LNCS. 2011, pp. 362–375. DOI: 10.1007/978-3-642-20807-2_29.

[PT88]   E. R. Petersen and A. J. Taylor. "An Optimal Scheduling System for the Welland Canal". In: *Transportation Sci.* 22.3 (1988), pp. 173–185.

[Phi+95]   C. Phillips, C. Stein, and J. Wein. "Scheduling jobs that arrive over time". In: *Proc. 4th International Workshop on Algorithms and Data Structures (WADS)*. Vol. 955. LNCS. 1995, pp. 86–97. DOI: 10.1007/3-540-60220-8_53.

[Phi+98]   C. Phillips, C. Stein, and J. Wein. "Minimizing average completion time in the presence of release dates". In: *Math. Program.* 82 (1998), pp. 199–223. DOI: 10.1007/BF01585872.

[PK00]   C. N. Potts and M. Y. Kovalyov. "Scheduling with batching: A review". In: *European J. Oper. Res.* 120.2 (2000), pp. 228–249.

[QS00]   M. Queyranne and M. Sviridenko. "New and improved algorithms for minsum shop scheduling". In: *Proc. 11th Annual ACM-SIAM Symposium on Discrete Algorithms (SODA)*. SIAM, 2000, pp. 871–878.

[RS10]   J. Rambau and C. Schwarz. *How to avoid collisions in scheduling industrial robots?* Preprint. opus.ub.uni-bayreuth.de/volltexte/2010/735/. Universität Bayreuth, 2010.

[Rig14]   G. Righini. "A network flow model of the Northern Italy waterway system". In: *EURO Journal on Transportation and Logistics* (2014), pp. 1–24. DOI: 10.1007/s13676-014-0068-y.

[Sch98]   C. Scheideler. "Offline routing protocols". In: *Universal Routing Strategies for Interconnection Networks*. Vol. 1390. LNCS. 1998, pp. 57–71.

[Sch11]   T. Schlechte. "Railway Track Allocation - Models and Algorithms". PhD thesis. TU Berlin, 2011.

[ST98]   P. Schonfeld and C.-J. Ting. "Optimization Through Simulation of Waterway Transportation Investments". In: *Transp. Res. Rec.* 1620 (1998), pp. 11–16.

[SW05]   P. Schonfeld and S.-L. Wang. "Scheduling Interdependent Waterway Projects through Simulation and Genetic Optimization". In: *J. Waterw. Port Coastal Ocean Eng.* 131.3 (2005), pp. 89–97. DOI: 10.1061/(ASCE)0733-950X(2005)131:3(89).

[Sch68]   L. Schrage. "A Proof of the Optimality of the Shortest Remaining Processing Time Discipline". In: *Oper. Res.* 16 (1968), pp. 687–690.

[SS02a]   A. S. Schulz and M. Skutella. "Scheduling Unrelated Machines by Randomized Rounding". In: *SIAM J. Discrete Math.* 15.4 (2002), pp. 450–469. DOI: 10.1137/S0895480199357078.

[SS02b]   A. S. Schulz and M. Skutella. "The Power of α-Points in Preemptive Single Machine Scheduling". In: *J. Sched.* 5 (2002), pp. 121–133. DOI: 10.1002/jos.93.

[Sei00]   S. S. Seiden. "A guessing game and randomized online algorithms". In: *Proc. 32nd Annual ACM Symposium on the Theory of Computing (STOC)*. 2000, pp. 592–601. DOI: 10.1145/335305.335385.

[Sga98]   J. Sgall. "On-line scheduling". In: *Online Algorithms: The State of the Art*. Ed. by A. Fiat and G. Woeginger. Vol. 1442. LNCS. Springer Berlin Heidelberg, 1998, pp. 196–231. DOI: 10.1007/BFb0029570.

[Sit10a]   R. Sitters. "Competitive analysis of preemptive single-machine scheduling". In: *Oper. Res. Lett.* 38 (2010), pp. 585–588. DOI: 10.1016/j.orl.2010.08.012.

[Sit10b]   R. Sitters. "Efficient Algorithms for Average Completion Time Scheduling". In: *Proc. 14th International Conference on Integer Programming and Combinatorial Optimization, (IPCO)*. Vol. 6080. LNCS. 2010, pp. 411–423. DOI: 10.1007/978-3-642-13036-6_31.

[Sku09]   M. Skutella. "An Introduction to Network Flows Over Time". In: *Research Trends in Combinatorial Optimization*. Ed. by W. Cook, L. Lovász, and J. Vygen. Berlin, Heidelberg: Springer, 2009, pp. 451–482.

[SW11]   M. Skutella and W. Welz. "Route Planning for Robot Systems". In: *Operations Research Proceedings 2010*. Ed. by B. Hu, K. Morasch, S. Pickl, and M. Siegle. Springer Berlin Heidelberg, 2011, pp. 307–312. DOI: 10.1007/978-3-642-20009-0_49.

[ST85]   D. D. Sleator and R. E. Tarjan. "Amortized Efficiency of List Update and Paging Rules". In: *Commun. ACM* 28 (1985), pp. 202–208. DOI: 10.1145/2786.2793.

[Smi56]   W. E. Smith. "Various optimizers for single-stage production". In: *Naval Research Logistics Quarterly* 3 (1956), pp. 59–66.

[SD88]   M. M. Solomon and J. Desrosiers. "Survey Paper—Time Window Constrained Routing and Scheduling Problems". In: *Transportation Sci.* 22.1 (1988), pp. 1–13.

[Ste08]   B. Stenzel. "Online Disjoint Vehicle Routing with Application to AGV Routing". PhD thesis. TU Berlin, 2008.

[SV02]   L. Stougie and A. P. A. Vestjens. "Randomized algorithms for on-line scheduling problems: How low can't you go?" In: *Oper. Res. Lett.* 30 (2002), pp. 89–96. DOI: 10.1016/S0167-6377(01)00115-8.

[Szp73]   B. Szpigel. "Optimal train scheduling on a single track railway". In: *Operational Research '72*. Ed. by M. Ross. North-Holland, Amsterdam, 1973, pp. 343–352.

[Tov84]   C. A. Tovey. "A simplified NP-complete satisfiability problem". In: *Discrete Appl. Math.* 8.1 (1984), pp. 85 –89.

[Vaz01]   V. V. Vazirani. *Approximation algorithms*. Springer Science & Business Media, 2001.

[VVB09]   J. Verstichel and G. Vanden Berghe. "A Late Acceptance Algorithm for the Lock Scheduling Problem". In: *Logistik Management*. Ed. by S. Voß, J. Pahl, and S. Schwarze. Physica-Verlag HD, 2009, pp. 457–478. DOI: 10.1007/978-3-7908-2362-2_23.

[Ves97]   A. P. A. Vestjens. "On-line Machine Scheduling". PhD thesis. Eindhoven University of Technology, Netherlands, 1997.

[WN14]   L. A. Wolsey and G. L. Nemhauser. *Integer and combinatorial optimization*. John Wiley & Sons, 2014.

List of Figures

List of Tables

List of Algorithms